Operation-Based Multi-Queue SBC Process Algebra For Systems Definition

-- Integration of Systems Structure and Systems Behavior --

Second Edition

William S. Chao

Structure-Behavior Coalescence

Systems Architecture = Systems Structure + Systems Behavior

CONTENTS

CONTENTS 5
PREFACE TO THE SECOND EDITION 9
PREFACE TO THE FIRST EDITION 11
ABOUT THE AUTHOR 13
PART I: BASIC CONCEPTS 15
   Chapter 1: Introduction to Systems Definition 17
      1-1 Physical and Virtual Systems 17
      1-2 Boundary and Environment of a System 19
      1-3 Systems Definition 1.0 21
      1-4 Systems Definition 2.0 23
   Chapter 2: Introduction to Process Algebra 25
      2-1 What is Process Algebra? 25
      2-2 Examples of Process Algebras 25
      2-3 Generalized SBC Process Algebra 25
      2-4 Specialized SBC Process Algebras 26
   Chapter 3: Mathematics of SBC Process Algebra 27
      3-1 Sequentialization of Prefixes 27
      3-2 Summation of Processes 27
      3-3 Parallel Composition of Processes 27
      3-4 Recursive Definition of a Process 28
      3-5 Replication of a Process 28
      3-6 Conditional Definition of a Process 28
      3-7 Null Process 28
PART II: OPERATION-BASED MULTI-QUEUE SBC PROCESS ALGEBRA 31
   Chapter 4: Operation-Based Value-Passing Interactions 33
      4-1 Operations 33
      4-2 Operation-Based Interactions 34
      4-3 Formal Description of an Operation-Based Communication Port.........39
      4-4 Formal Description of an Operation-Based Action 40
      4-5 Formal Description of an Operation-Based Interaction.....................41
   Chapter 5: The Structure-Behavior Coalescence Approach 43
      5-1 Structure-Behavior Coalescence Means to Integrate the Systems Structure and Systems Behavior 43
      5-2 Interactions among Components and the External Environment to Draw

  Forth the Systems Behavior 44

  5-3 Core Theme of Structure-Behavior Coalescence..............................46

Chapter 6: Language Constructs of Operation-Based Multi-Queue SBC Process Algebra 49

  6-1 Entity Set and Entity Name 49

  6-2 Examples of Entities 50

  6-3 Backus-Naur Form of Operation-Based Multi-Queue SBC Processes .54

    6-3-1 Parallel Composition of One or More Recursive Interaction Flow Diagrams Defines the Operation-Based SBC Process of a System 54

    6-3-2 A Recursive Interaction Flow Diagram is the Recursion of an Interaction Flow Diagram 55

    6-3-3 An Interaction Flow Diagram is a Type_1 Expression Followed by Zero or More Type_1_Or_Type_2 Expressions 55

    6-3-4 Type_1_Or_2 Expression is either Type_1 or Type_2 56

    6-3-5 Summation of One or More Condition Type_1 Interactions Defines the Type_1 Expression 56

    6-3-6 Summation of One or More Condition Type_2 Interactions Defines the Type_2 Expression 57

  6-4 Examples of Type_1 Expression 57

  6-5 Examples of Type_2 Expression 65

Chapter 7: Transitional Semantics of Operation-Based Multi-Queue SBC Process Algebra 75

  7-1 Transitional Semantics...75

  7-2 Rule of Prefix..76

  7-3 Rule of Summation...77

  7-4 Rule of Parallel Composition..77

  7-5 Rule of Recursion 78

  7-6 Rule of Constants..79

PART III: CASES STUDY 81

 Chapter 8: Operation-Based Multi-Queue SBC Process of the Restaurant.83

  8-1 BNF Tree of the Restaurant 83

  8-2 Prefixes of the Restaurant 86

  8-3 Process of the Restaurant 92

 Chapter 9: Operation-Based Multi-Queue SBC Process of the Car 95

  9-1 BNF Tree of the Car..95

  9-2 Prefixes of the Car 98

  9-3 Process of the Car 102

APPENDIX A: Language Constructs of Operation-Based Multi-

Queue SBC Process Algebra 105
APPENDIX B: Transitional Semantics of Operation-Based Multi-Queue SBC Process Algebra 107
BIBLIOGRAPHY 109
INDEX 113

PREFACE TO THE SECOND EDITION

In the first edition of this book, we do not treat any conditional expression. Therefore, all the theories of operation-based multi-queue SBC process algebra (O-M-SBC-PA) attempt to ignore all conditional expressions.

After further research, we find that if we embed each condition into a prefix, all the theories of operation-based multi-queue SBC process algebra will be more complete and beautiful.

In the second edition of this book, the reader would enjoy to probe conditional expressions that are so elegantly handled in every aspect of operation-based multi-queue SBC process algebra.

PREFACE TO THE FIRST EDITION

The need for systems definition arises because any real-life system is inherently complicated. It is impossible to comprehend fully the intricate interaction of any system of the real world with its environment, or to define all its components and each of its details. Systems definition or system definition is an artifact created by humans to define what a system is.

Process algebras are a diverse family of related approaches to the study of concurrent systems. Their tools are algebraic languages for the high-level description of interactions, communications and synchronizations between a collection of independent agents or processes. Process algebras also provide algebraic laws that allow process descriptions to be manipulated and analyzed, and permit formal reasoning about equivalences and observation congruence among processes. Accordingly, process algebra provides a perfect method for systems definition.

Operation-based multi-queue SBC process algebra (O-M-SBC-PA) is one of the six specialized SBC process algebras. In this book, we use O-M-SBC-PA to achieve the robust systems definition of a system. To see is to believe. Therefore, many examples are presented to help the reader fully understand the use of O-M-SBC-PA.

ABOUT THE AUTHOR

Dr. William S. Chao is the CEO & founder of SBC Architecture International®. SBC (Structure-Behavior Coalescence) architecture is a systems architecture which demands the integration of systems structure and systems behavior of a system. SBC architecture applies to hardware architecture, software architecture, enterprise architecture, knowledge architecture and thinking architecture. The core theme of SBC architecture is: "Architecture = Structure + Behavior."

William S. Chao received his bachelor degree (1976) in telecommunication engineering and master degree (1981) in information engineering, both from the National Chiao-Tung University, Taiwan. From 1976 till 1983, he worked as an engineer at Chung-Hwa Telecommunication Company, Taiwan.

William S. Chao received his master degree (1985) in information science and Ph.D. degree (1988) in information science, both from the University of Alabama at Birmingham, USA. From 1988 till 1991, he worked as a computer scientist at GE Research and Development Center, Schenectady, New York, USA.

Dr. William S. Chao has been teaching at National Sun Yat-Sen University, Taiwan since 1992 and now serves as the president of Association of Enterprise Architects, Taiwan Chapter. His research covers: systems architecture, hardware architecture, software architecture, enterprise architecture, knowledge architecture and thinking architecture.

PART I: BASIC CONCEPTS

Chapter 1: Introduction to Systems Definition

All things that amaze us as something independent are essentially parts of a system. We usually call the parts of a system its components. Every system is something the whole. Systems emphasize the holistic vision.

The need for systems definition arises because any real-life system is inherently complicated. It is impossible to comprehend fully the intricate interaction of any system of the real world with its environment, or to define all its components and each of its details. Systems definition or system definition is an artifact created by humans to define what a system is.

In this chapter, we first introduce physical and conceptual systems. A physical system exists in the physical, concrete, or real world. A conceptual system exists in the conceptual, abstract, or virtual world. We then introduce the boundary of a system. A system has a boundary. The system itself is inside the boundary and the environment is outside the boundary. After these introductions, we shall thereafter discuss the systems definition 1.0 as well as the systems definition 2.0.

1-1 Physical and Virtual Systems

In general, systems are divided into two categories: 1) physical systems and 2) virtual systems.

A physical system exists in the physical world [Acko68]. A physical system is also called a concrete or real system. For example, a *telephone* composed of *microphone, earphone and keypad,* shown in Figure 1-1, is a physical, concrete, or real system.

Figure 1-1   A Telephone is a Physical System

As a second example, a *stool* composed of *seat* and *legs*, shown in Figure 1-2, is a physical, concrete, or real system.

Figure 1-2   A Stool is a Physical System

A virtual system is a system that is composed of non-physical components, i.e., ideas, thoughts, or notions. A virtual system exists in the virtual, abstract, or notional world. For example, the "*Snow White and the Seven Dwarfs*" fairy tale composed of "*Snow White*" and "*Seven Dwarfs*," shown in Figure 1-3, is a virtual, abstract, or notional system.

Figure 1-3 *Snow White and the Seven Dwarfs* is a Virtual System

As a second example, the "*Multi-Tier Personal Data System*" software composed of *MTPDS_GUI, Age_Logic, Overweight_Logic* and *Personal_Database*, shown in Figure 1-4, is a virtual, abstract, or notional system.

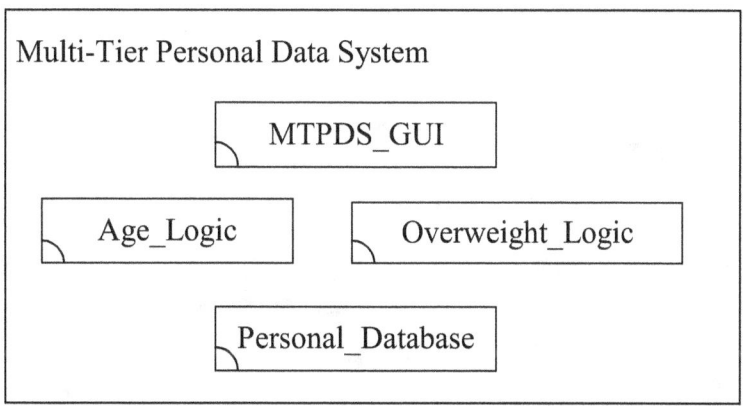

Figure 1-4 *Multi-Tier Personal Data System* is a Virtual System

1-2 Boundary and Environment of a System

We scope a system by describing its boundary as shown in Figure 1-5. All components of the system are inside the boundary while the environment is outside the boundary.

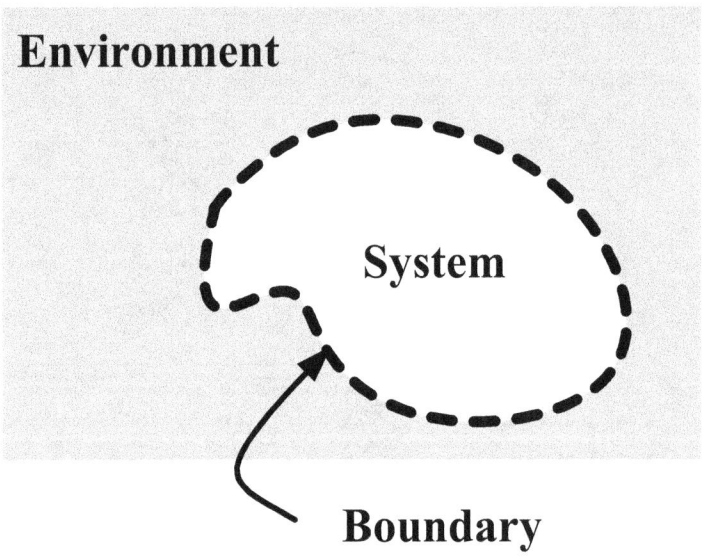

Figure 1-5 Boundary and Environment of a System

The environment is also known as the surroundings. A system may or may not interrelate with the environment. An open system interrelates with the environment through the exchange of matter, energy, data, information, or message as shown in Figure 1-6.

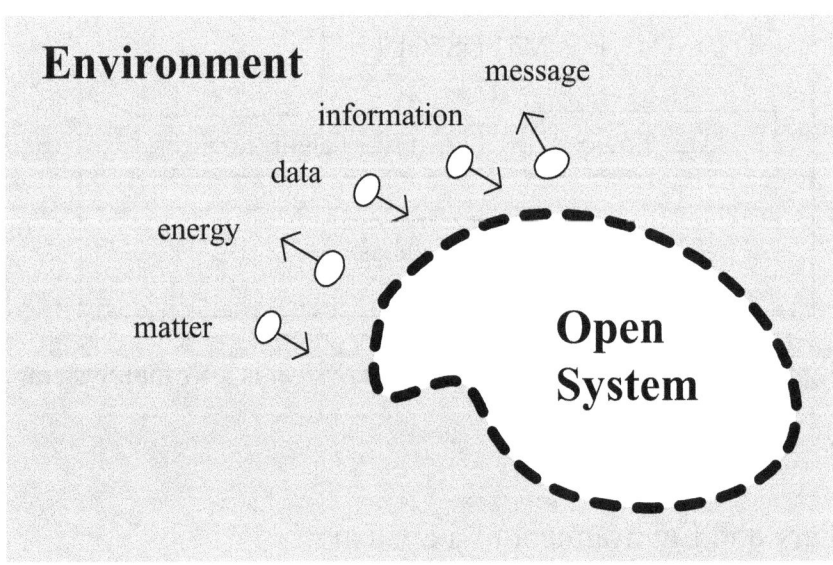

Figure 1-6 Open System Interrelates with the Environment

An isolated system does not interrelate with the environment at all. There is no exchange of matter, energy, data, information, or message between the isolated system and the environment as shown in Figure 1-7.

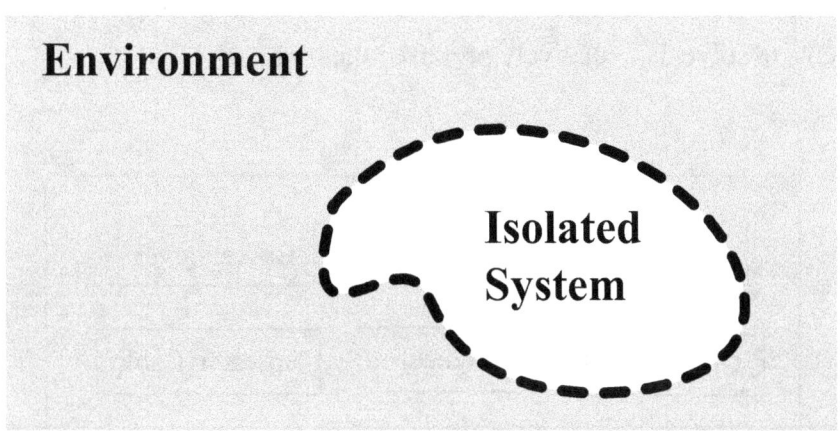

Figure 1-7   Isolated System Does Not Interrelate with the Environment

1-3 Systems Definition 1.0

Systems definition 1.0 defines a system, in Figure 1-8, hopefully to be an integrated whole, embodied in its assembled components, their interrelationships with each other and the environment [Chec99, Frie11, Ghar11, Mead08].

> A system, hopefully is an integrated whole,
> embodied in its assembled components,
> their interrelationships with each other and the environment.

Figure 1-8   Systems Definition 1.0 Defining a System

Components are sometimes labeled as parts, entities, objects, building blocks and non-aggregated systems [Chao14a, Chao14b, Chao14c]. Interrelated components make a system not only a whole but also hopefully an integrated whole.

A system defined by systems definition 1.0 has the following characteristics: 1) hopefully, it is an integrated whole; 2) it is embodied in its assembled components; 3) components are interrelated with each other and the environment; and 4) it uses structural decomposition [Chao14b, Ghar11] rather than functional decomposition [Scho10].

The structural decomposition method is to decompose a system into a number of components, as shown in Figure 1-9. Breaking down a large problem into a number of components to solve, is a relatively preferred method.

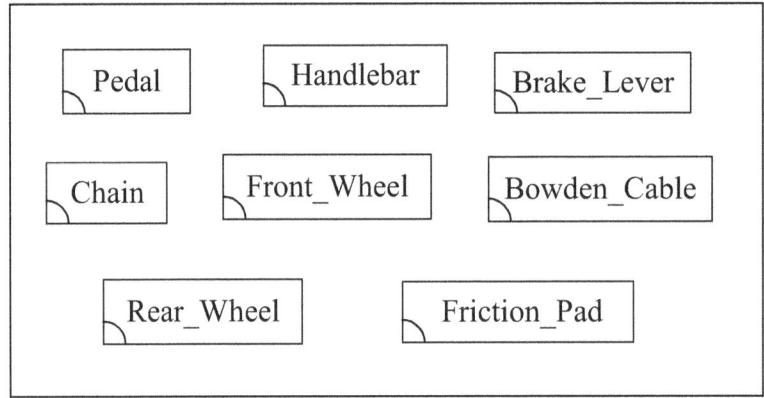

Figure 1-9 Structural Decomposition Method

The functional decomposition method is to decompose a system into a number of functions, as shown in Figure 1-10. Breaking down a large problem into a number of functions to solve, is a relatively non-preferred method.

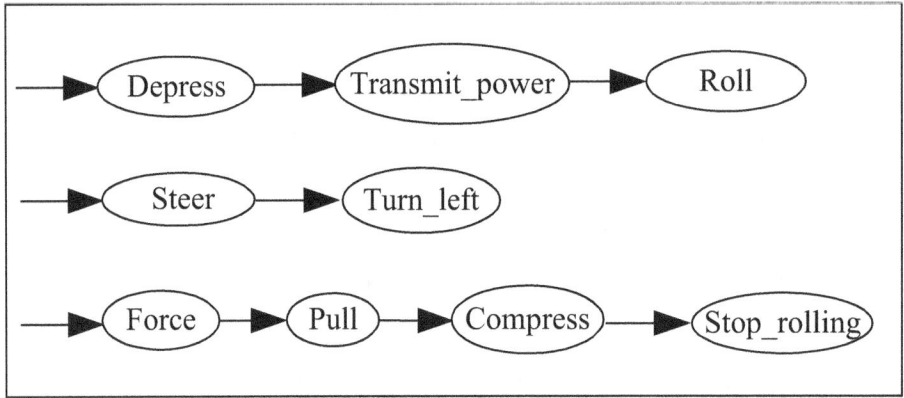

Figure 1-10 Functional Decomposition Method

1-4 Systems Definition 2.0

Systems structure and systems behavior are the two most significant views of a system. In order to achieve a truly integrated whole of a system, we first need to integrate the systems structure and systems behavior together. In other words, integration of systems structure and systems behavior results in the integration of a whole system. Since systems definition 1.0 does not define the integration of systems structure and systems behavior, very likely it only hopes and will never be able to really form an integrated whole of a system. In this situation, systems definition 1.0 is powerless in defining a system appropriately.

Structure-behavior coalescence (SBC) provides an elegant way to integrate the systems structure and systems behavior of a system. A system is therefore redefined, by systems definition 2.0, truly to be an integrated whole, through structure-behavior coalescence, embodied in its assembled components, their interactions with each other and the environment as shown in Figure 1-11.

A system,
through the SBC approach,
truly is an integrated whole,
embodied in its assembled components,
their interactions with each other and the environment.

Figure 1-11　Systems Definition 2.0 Defining a System

Systems definition or system definition 2.0 formally defines the essence of a system and its details at the same time. Since systems definition 2.0 demands the integration of systems structure and systems behavior, definitely it is able to form an integrated whole of a system. In this situation, systems definition 2.0 is fully capable of defining a system.

A system defined by systems definition 2.0 has the following characteristics: 1) it emphasizes the system's structure-behavior coalescence; 2) it is a truly integrated whole; 3) it is embodied in its assembled components; 4) components are interacting (or handshaking) with each other and the environment; and 5) it uses structural decomposition [Chao14a, Chao14b, Chao14c, Ghar11] rather than functional decomposition [Scho10].

Chapter 2: Introduction to Process Algebra

In this chapter, we first discuss what process algebra is. Then we show several examples of process algebras. After that we elaborate on the generalized SBC process algebra and six specialized SBC process algebras.

2-1 What is Process Algebra?

Process algebras are a diverse family of related approaches to the study of concurrent systems [Berg87, Chao15d, Hoar85, Miln89, Miln99]. Their tools are algebraic languages for the high-level description of interactions, communications and synchronizations between a collection of independent agents or processes.

Process algebras also provide algebraic laws that allow process descriptions to be manipulated and analyzed, and permit formal reasoning about equivalences and observation congruence among processes.

2-2 Examples of Process Algebras

There are several leading algebraic approaches to modeling concurrent systems.

Communicating Sequential Processes (CSP) [Hoar85] was first described in a 1978 paper by C. A. R. Hoare.

Arthur John Robin Gorell Milner introduced the Calculus of Communicating Systems (CCS) [Miln89, Miln99] around 1980.

Algebra of Communicating Processes (ACP) [Berg87] was initially developed by Jan Bergstra and Jan Willem Klop in 1982.

2-3 Generalized SBC Process Algebra

Generalized SBC process algebra (G-SBC-PA) evolved from CCS (Calculus of Communicating Systems) [Miln89, Miln99].

CCS is a general process algebra language for the study of communication and concurrency. Like CCS, generalized SBC process algebra is also a general process algebra language for the study of communication and concurrency.

2-4 Specialized SBC Process Algebras

Channel-based single-queue SBC process algebra (C-S-SBC-PA) [Chao15d, Chao15e, Chao15g], channel-based multi-queue SBC process algebra (C-M-SBC-PA) [Chao15d, Chao15f, Chao15h], channel-based infinite-queue SBC process algebra (C-I-SBC-PA) [Chao15b, Chao15c, Chao15d], operation-based single-queue SBC process algebra (O-S-SBC-PA), operation-based multi-queue SBC process algebra (O-M-SBC-PA) [Chao15d, Chao15f, Chao15h] and operation-based infinite-queue SBC process algebra (O-I-SBC-PA) [Chao15b, Chao15c, Chao15d] are the six specialized SBC process algebras.

Channel-based single-queue SBC process algebra, channel-based multi-queue SBC process algebra, channel-based infinite-queue SBC process algebra, operation-based single-queue SBC process algebra, operation-based multi-queue SBC process algebra and operation-based infinite-queue SBC process algebra all evolved from CCS (Calculus of Communicating Systems) [Miln89, Miln99].

CCS is a general process algebra language for the study of concurrent systems. Unlike CCS, six specialized SBC process algebras are only applicable to systems definition [Burd10, Maie09, Chao16b, Chao16c, Chao16d, Chao16e, Chao16f, Chao16g, Craw15, Dam06, O'Rou03, Putm00, Rayn09, Roza11, Toga08].

Chapter 3: Mathematics of SBC Process Algebra

To give the SBC process a mathematical definition, we need a means to form new processes from old ones. The basic operators, always present in some form or other, allow sequentialization of prefixes or summation of processes or parallel composition of processes or recursive definition of a process or replication of processes or conditional definition of a process or null process.

3-1 Sequentialization of Prefixes

Sometimes prefixes must be temporally ordered. For example, it might be desirable to specify algorithms such as: execute the "t" prefix (interaction with a condition) first and then execute the "P" process later. Sequentialization of prefixes can be used for such purposes.

Sequentialization of prefixes, usually written as the $t \bullet P$ process, indicates that it will perform the "t" prefix first and continue as the "P" process.

3-2 Summation of Processes

The binary operator "+", summation, combines two process expressions as alternatives.

For example, the process $P+Q$ can proceed non-deterministically either as the process P or the process Q; as soon as one performs its first interaction the other is discarded.

3-3 Parallel Composition of Processes

Parallel composition of two processes P and Q, usually written $P\|Q$, is the key primitive distinguishing the process algebras from sequential models of process executions.

Parallel composition allows the executions in P and Q to proceed simultaneously and independently.

3-4 Recursive Definition of a Process

The operators presented so far describe only finite interaction and are consequently insufficient for full computability, which includes non-terminating behavior. Recursion is the operator that allows finite descriptions of infinite behavior.

For example, **fix**($X=E$) can be understood as abbreviating the recursive definition of an infinite behavior denoted by the "X" process variable.

3-5 Replication of a Process

Replication is the other operator that allows finite descriptions of infinite behavior of a process.

For example, replication $!P$ can be understood as abbreviating the parallel composition of a countably infinite number of P processes.

3-6 Conditional Definition of a Process

A process can be defined by a one-or-more-armed conditional expression. For example, the process (**if** $cond_1$ **then** a_1)+(**if** $cond_2$ **then** a_2)…+(**if** $cond_j$ **then** a_j) will proceed to perform the "a_1" interaction if the "$cond_1$" value is true, or proceed to perform the "a_2" interaction if the "$cond_2$" value is true,…, or proceed to perform the "a_j" interaction if the "$cond_j$" value is true.

3-7 Null Process

Process algebras generally also include a null process, denoted as *STOP*, which has no interaction points. It is utterly inactive and its sole purpose is to act as the inductive anchor on top of which more interesting processes can be generated.

The process "*STOP*•P_1" (i.e. sequential composition of processes *STOP* and P_1) equals to the process "*STOP*", as shown in Figure 3-1.

$$STOP \bullet P_1 = STOP$$

Figure 3-1   Characteristics of Null Process (I)

The process "P_2+STOP" (i.e. summation of processes P_2 and $STOP$) equals to the process "$STOP+P_2$" (i.e. summation of processes $STOP$ and P_2) which equals to the process "P_2", as shown in Figure 3-2.

$$P_2 + STOP = STOP + P_2 = P_2$$

Figure 3-2   Characteristics of Null Process (II)

The process "$P_3 \| STOP$" (i.e. parallel composition of processes P_3 and $STOP$) equals to the process "$STOP \| P_3$" (i.e. parallel composition of processes $STOP$ and P_3) which equals to the process "P_3", as shown in Figure 3-3.

$$P_3 \| STOP = STOP \| P_3 = P_3$$

Figure 3-3   Characteristics of Null Process (III)

PART II: OPERATION-BASED MULTI-QUEUE SBC PROCESS ALGEBRA

Chapter 4: Operation-Based Value-Passing Interactions

In this chapter, we first introduce operations and operation-based interactions. We then introduce the formal description of an operation-based communication port, an operation-based action and an operation-based interaction.

4-1 Operations

An operation provided by each component represents a procedure, or method, or function of the component as shown in Figure 4-1.

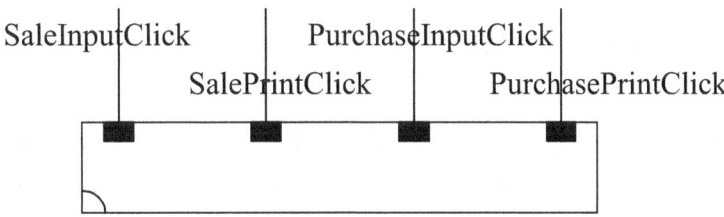

Figure 4-1   An Operation Represents a Procedure, or Method, or Function of a Component

An operation may contain several input parameters (e.g. i_1, i_2) and output parameters (e.g. o_1, o_2), as shown in Figure 4-2.

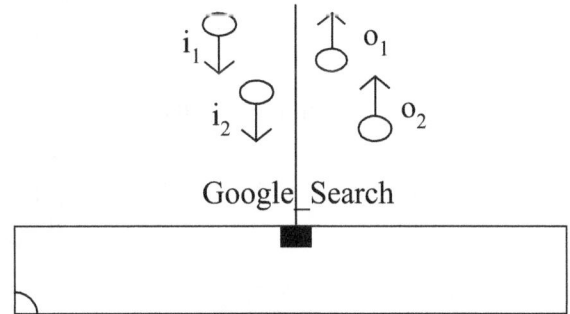

Figure 4-2   An Operation Contains Several Input/Output Parameters

An operation formula is used to completely describe an operation. An operation formula includes a) operation name, b) input parameters (e.g. $i_1, i_2, ..., i_m$) and c) output parameters (e.g. $o_1, o_2, ..., o_n$), as shown in Figure 4-3.

$$\text{Operation\_Name (In } i_1, i_2, ..., i_m \text{ ; Out } o_1, o_2, ..., o_n \text{)}$$

Figure 4-3   Operation Formula

4-2 Operation-Based Interactions

An interaction represents an indivisible and instantaneous handshake or communication between two agents. In the operation-based approach as shown in Figure 4-4, the caller agent (either external environment's actor or component) interacts with the callee agent (component) through the operation call or operation return interaction.

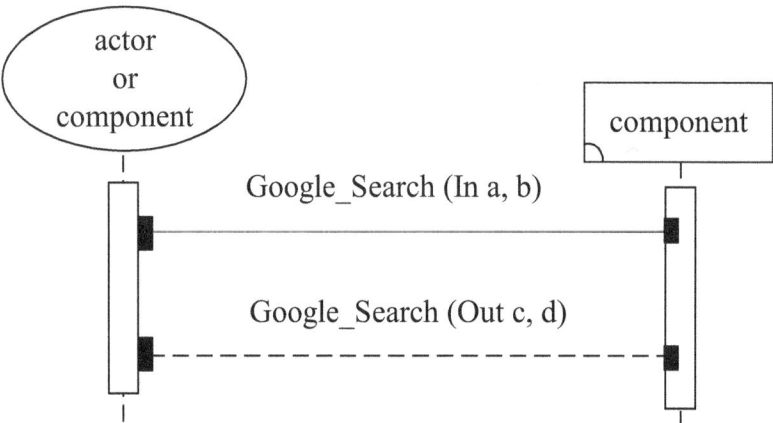

Figure 4-4   Operation-Based Interactions

The caller agent owns the "calling port" of the interaction. In the operation call interaction case, the calling port is "Google_Search (In a, b)" whose conduct is to output a value to each one of the "a" and "b" variables (of the "Google_Search" operation) as shown in Figure 4-5.

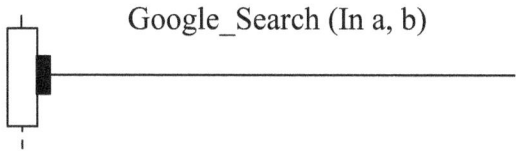

Figure 4-5   Calling Port
in the Operation Call Interaction Case

The caller agent together with the "calling port" is named the "calling action" as shown in Figure 4-6.

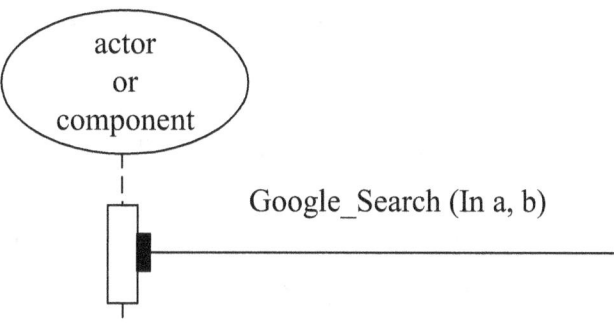

Figure 4-6   Calling Action
in the Operation Call Interaction Case

In the operation return interaction case, the calling port is "Google_Search (Out c, d)" whose conduct is to receive a value from each one of the "c" and "d" variables (of the "Google_Search" operation) as shown in Figure 4-7.

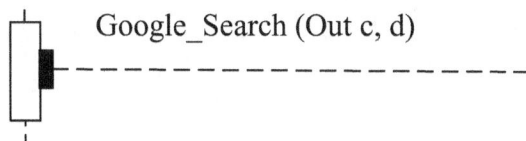

Figure 4-7   Calling Port
in the Operation Return Interaction Case

The caller agent together with the "calling port" is named the "calling action" as shown in Figure 4-8.

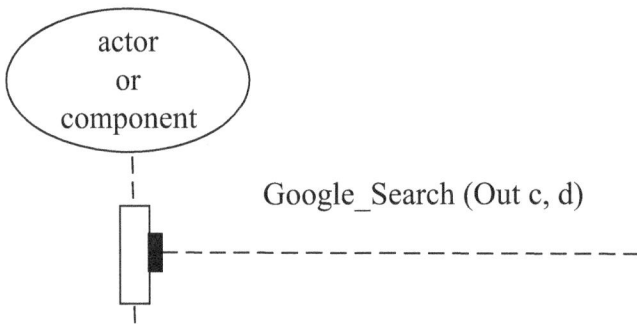

Figure 4-8   Calling Action
in the Operation Return Interaction Case

The callee component owns the "called port" of the interaction. In the operation call interaction case, the called port is "Google_Search (In a, b)" whose conduct is (using the "Google_Search" operation) to receive a value for each one of the "a" and "b" variables as shown in Figure 4-9.

Figure 4-9   Called Port
in the Operation Call Interaction Case

The callee component together with the "called port" is named the "called action" as shown in Figure 4-10.

Figure 4-10   Called Action
in the Operation Call Interaction Case

In the operation return interaction case, the called port is "Google_Search (Out c, d)" whose conduct is (using the "Google_Search" operation) to output the values of the "c" and "d" variables as shown in Figure 4-11.

Figure 4-11   Called Port
in the Operation Return Interaction Case

The callee component together with the "called port" is named the "called action" as shown in Figure 4-12.

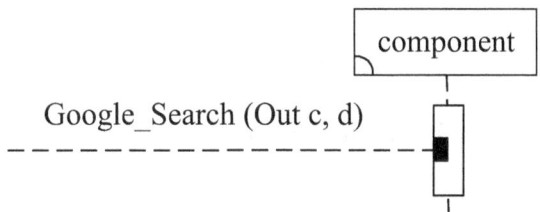

Figure 4-12   Called Action
in the Operation Return Interaction Case

In order to simplify the operation-based interaction diagram, we will redraw it as shown in Figure 4-13.

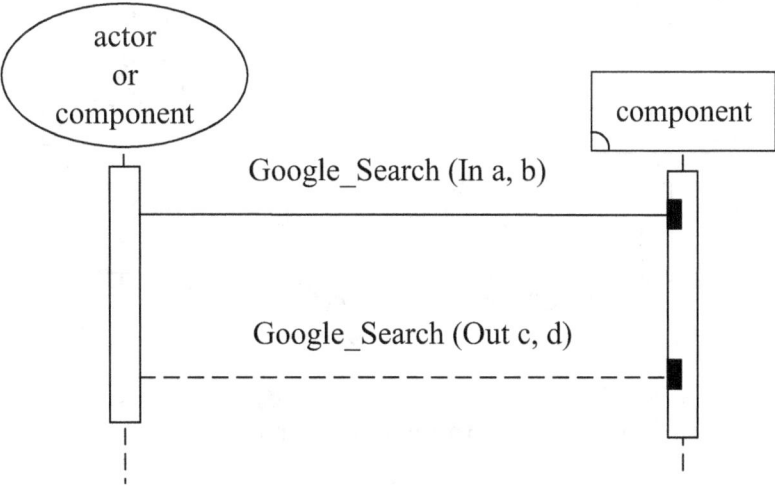

Figure 4-13   Operation-Based Interaction Diagram (I)

Or we can draw the operation-based interaction diagram as shown in Figure 4-14.

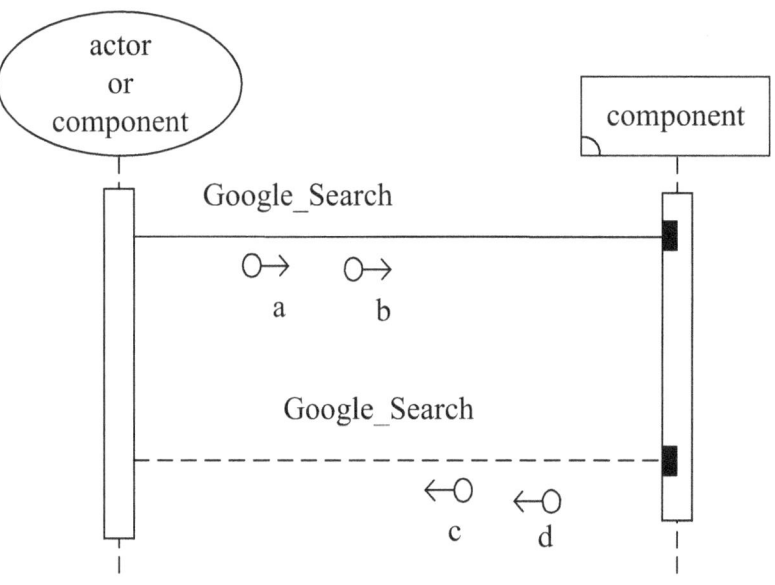

Figure 4-14   Operation-Based Interaction Diagram (II)

We use an operation-based internal interaction (i.e. λ) to represent their handshake or communication, if the caller component and the callee component are the same one, as shown in Figure 4-15.

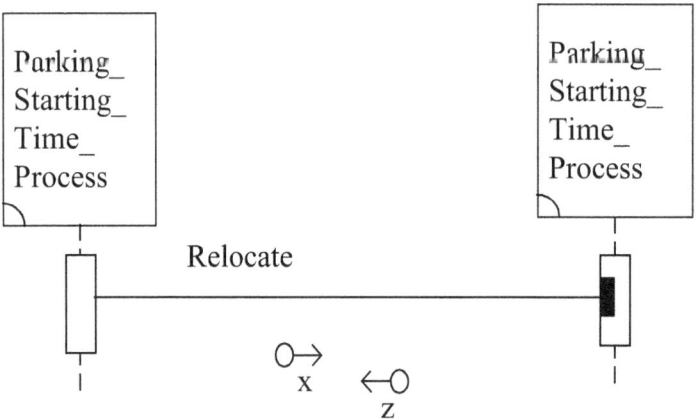

Figure 4-15   An Internal Interaction (I)

Also, we may redraw the internal interaction as shown in Figure 4-16.

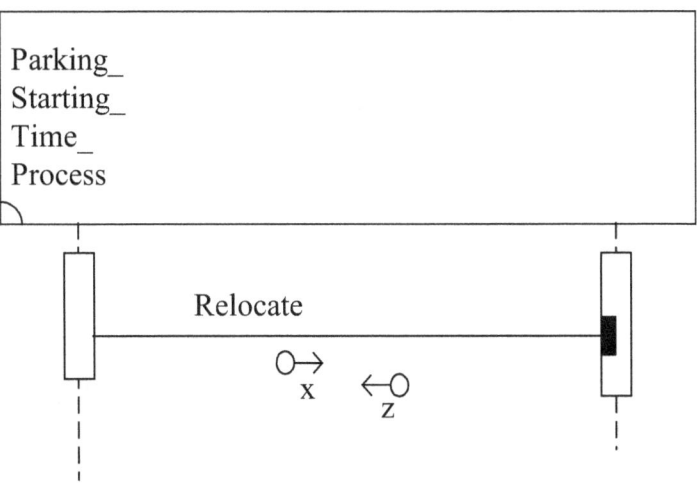

Figure 4-16　　An Internal Interaction (II)

4-3 Formal Description of an Operation-Based Communication Port

We formally describe an operation-based communication port as a 3-tuple PORT = <operation_call_or_return, calling_or_called, operation_call_or_return_formula>, where "operation_call_or_return" stands for an OPERATION_CALL or OPERATION_RETURN tag, "calling_or_called" stands for a CALLING or CALLED action tag, and "operation_call_or_return_formula" stands for an operation call or operation return formula as shown in Figure 4-17.

Figure 4-17 Formal Description of
an Operation-Based Communication Port

4-4 Formal Description of an Operation-Based Action

We formally describe an operation-based action as a 4-tuple ACTION = <operation_call_or_return, agent, calling_or_called, operation_call_or_return_formula>, where "operation_call_or_return" stands for an OPERATION_CALL or OPERATION_RETURN tag, "agent" stands for the name of a caller agent or callee component, "calling_or_called" stands for a CALLING or CALLED action tag, and "operation_call_or_return_formula" stands for an operation call or operation return formula as shown in Figure 4-18.

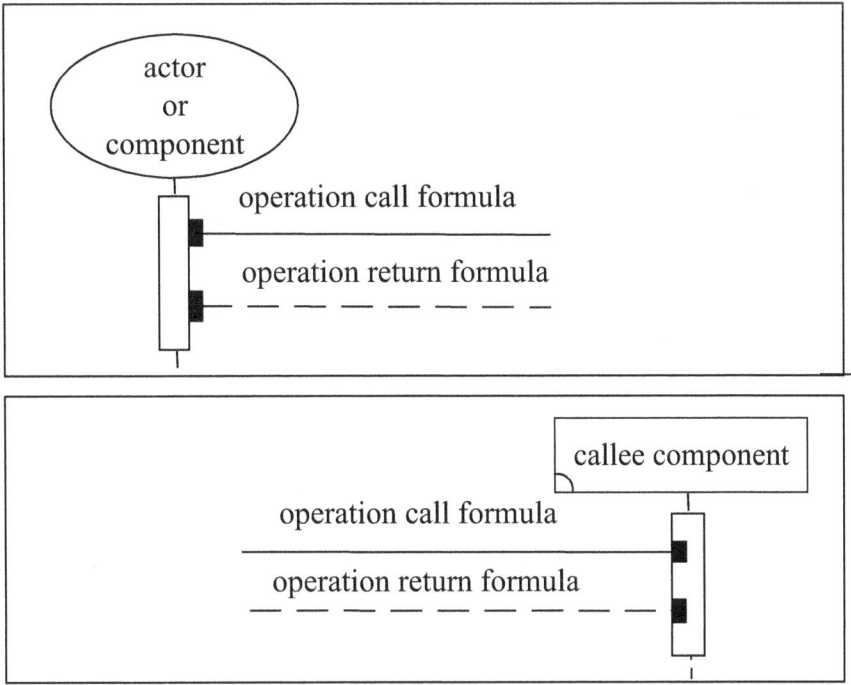

Figure 4-18   Formal Description of an Operation-Based Action

4-5 Formal Description of an Operation-Based Interaction

We formally describe an operation-based interaction as a 4-tuple INTERACTION = <operation_call_or_return, caller_agent, operation_call_or_return_formula, callee_component>, where "operation_call_or_return" stands for an OPERATION_CALL or OPERATION_RETURN tag, "caller_agent" stands for the name of an actor or a caller_component, "operation_call_or_return_formula" stands for an operation call or operation return formula and "callee_component" stands for the name of a callee component as shown in Figure 4-19.

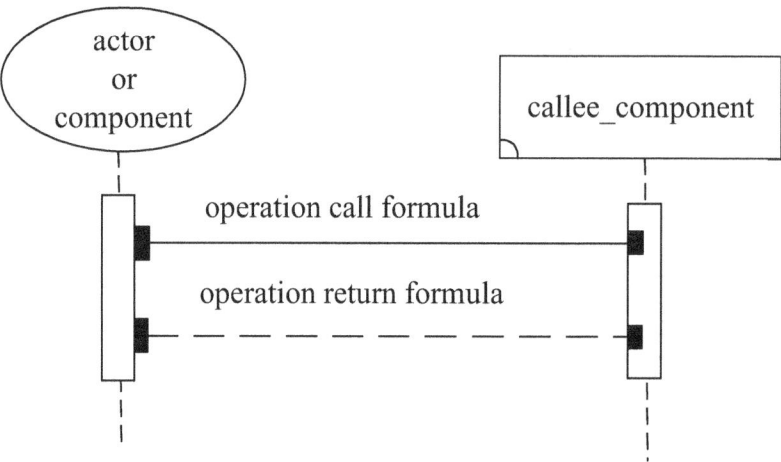

Figure 4-19 Formal Description of an Operation-Based Interaction

Chapter 5: The Structure-Behavior Coalescence Approach

Systems structure and systems behavior are the two most prominent views of a system, integrating the systems structure and systems behavior is apparently the best way to achieve an integrated whole of a system. If we are not able to integrate the systems structure and systems behavior, then there is no way that we are able to integrate the whole system. Structure-behavior coalescence (SBC) provides an elegant way to integrate the systems structure and systems behavior of a system. In other words, SBC facilitates an integrated whole of a system.

5-1 Structure-Behavior Coalescence Means to Integrate the Systems Structure and Systems Behavior

Systems structure, specified by components, their operations and their composition, refers to the type of connection between the components of a system as shown in Figure 5-1.

Figure 5-1   Systems Structure

Systems behavior, specified by the interactions between and among the components and environment, refers to the interconnectivities a system in conjunction with its environment as shown in Figure 5-2.

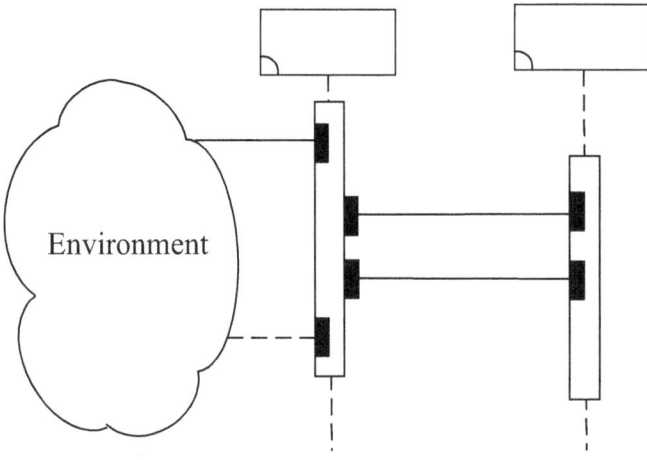

Figure 5-2   Systems Behavior

Systems structure and systems behavior are the two most prominent views of a system, integrating the systems structure and systems behavior is apparently the best way to achieve an integrated whole of a system.

If we are not able to integrate the systems structure and systems behavior, then there is no way that we are able to integrate the whole system.

Structure-behavior coalescence (SBC) [Chao15a] provides an elegant way to integrate the systems structure and systems behavior of a system. In other words, SBC facilitates an integrated whole of a system.

5-2 Interactions among Components and the External Environment to Draw Forth the Systems Behavior

All things that strike us as something independent are essentially parts of a system. We usually call the parts of a system its components. Components are sometimes labeled as parts, entities, objects, building blocks and non-aggregated systems [Chao14a, Chao14b, Chao14c, Chao16a].

In a system, if the components, and among them and the external environment to interact (or handshake), such interaction will draw forth the systems behavior.

The external environment uses a "type-1 interaction" to interact with a component as shown in Figure 5-3.

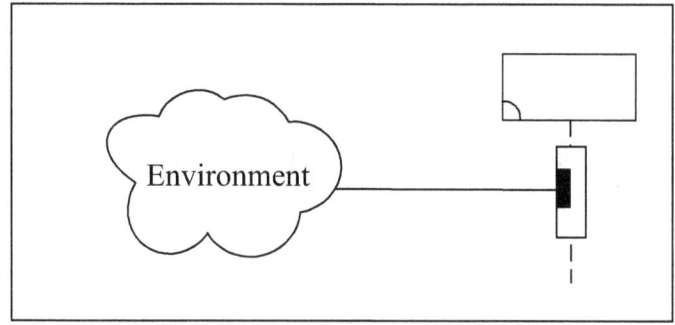

Figure 5-3   The External Environment Uses a "Type-1 Interction"
to Interact with a Component

We formally describe an operation-based "type-1 interaction" as a 4-tuple TYPE_1_INTERACTION = <operation_call_or_return, actor, operation_call_or_return_formula, callee_component>, where "operation_call_or_return" stands for an OPERATION_CALL or OPERATION_RETURN tag, "actor" stands for the name of an external environment's actor, "operation_call_or_return_formula" stands for an operation call or operation return formula and "callee_component" stands for the name of a callee component.

Two components use a "type-2 interaction" to interact with each other as shown in Figure 5-4.

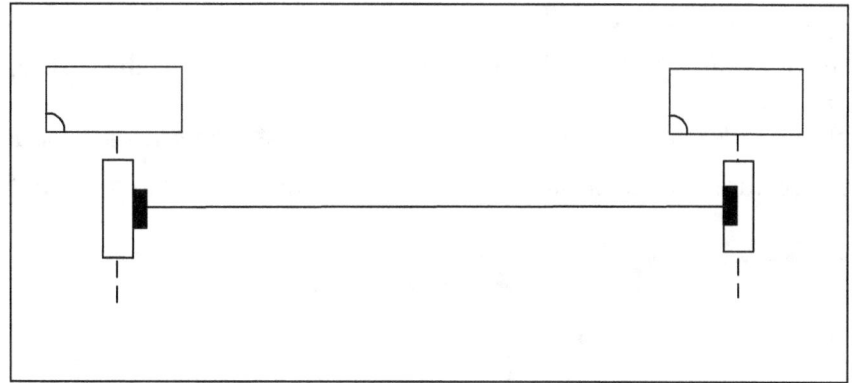

Figure 5-4   Two Components Use a "Type_2 Interaction"
to Interact with Each Other

We formally describe an operation-based "type-2 interaction" as a 4-tuple TYPE_2_INTERACTION = <operation_call_or_return, caller_component, operation_call_or_return_formula, callee_component>, where "operation_call_or_return" stands for an OPERATION_CALL or OPERATION_RETURN tag, "caller_component" stands for the name of a caller component, "operation_call_or_return_formula" stands for an operation call or operation return formula and "callee_component" stands for the name of a callee component.

5-3 Core Theme of Structure-Behavior Coalescence

The core theme of structure-behavior coalescence is: "Systems Architecture = Systems Structure + Systems Behavior." That is, the systems structure will lead to the systems behavior as shown in Figure 5-5.

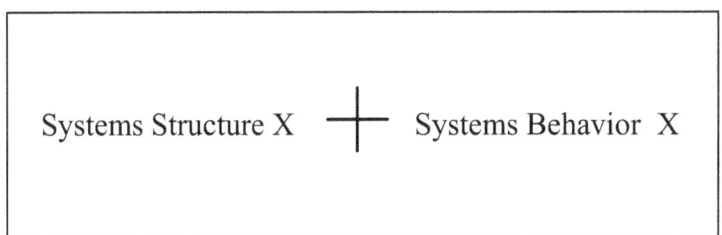

Figure 5-5   Core Theme of Structure-Behavior Coalescence

One systems structure will draw forth one systems behavior. That is, the systems behavior is attached to or built on the systems structure in the SBC approach.

In other words, the systems behavior can not exist alone; it must be loaded on the systems structure just like a cargo is loaded on a ship as shown in Figure 5-6.

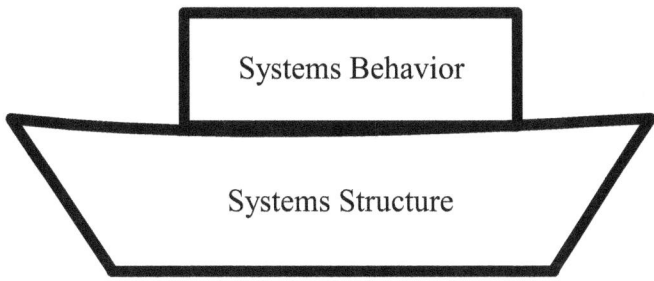

Figure 5-6 Systems Behavior Must be Loaded on the Systems Structure

Chapter 6: Language Constructs of Operation-Based Multi-Queue SBC Process Algebra

In the chapter, we illustrate in detail those operation-based multi-queue SBC process algebra language constructs which make up the system definition of a system.

6-1 Entity Set and Entity Name

As shown in Figure 6-1, we assume an infinite set L of operation call or operation return formulas, and use l to range over L. Further, we let N be the set of condition type_1 interactions, and use n to range over N. We let Z be the set of condition type_2 interactions, and use z to range over Z. We let T be the set of condition type_1_or_2 interactions, and use t to range over T. Further, we let X be the set of process variables, and use X to range over X. We let Φ be the set of process Constants, and use A, B to range over Φ. We let Π be the set of processes, and use P, Q to range over Π. We let Ψ be the set of process expressions, and use E, F to range over Ψ. Finally, we let Γ be the set of components, and use C, D to range over Γ.

Entity set	Entity name	Type of entity
L	$l,...$	operation call or operation return formulas
N	$n,...$	condition type_1 interactions
Z	$z,...$	condition type_2 interactions
T	$t,...$	condition type_1_or_2 interactions
Δ	$a,...$	type 1 or 2 interactions
X	$X,...$	process variables
Φ	$A, B,...$	process Constants
	$I, J,...$	indexing sets
Π	$P, Q,...$	processes
Ψ	$E, F,...$	process expressions
Γ	$C, D,...$	components

Figure 6-1　Entities

6-2 Examples of Entities

As a first example, the n_{001} prefix defined as "<OPERATION_CALL, $Actor_{001}$, l_{001}, C_{001}>" or "**if** *true* **then** <OPERATION_CALL, $Actor_{001}$, l_{001}, C_{001}>" is an operation-based condition type_1 interaction under a tautology condition in which "OPERATION_CALL" stands for an operation call interaction tag, "$Actor_{001}$" stands for the external environment, "l_{001}" stands for an operation call formula, and "C_{001}" stands for a callee component as shown in Figure 6-2.

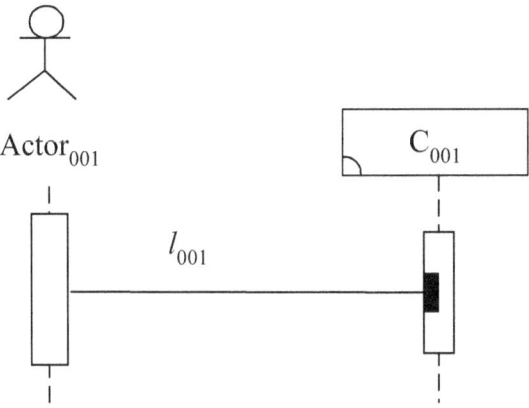

Figure 6-2　Prefix of n_{001}

As a second example, the n_{002} prefix defined as "<OPERATION_RETURN, $Actor_{002}$, l_{002}, C_{002}>" or "**if** *true* **then** <OPERATION_RETURN, $Actor_{002}$, l_{002}, C_{002}>" is an operation-based condition type_1 interaction under a tautology condition in which "OPERATION_RETURN" stands for an operation return interaction tag, "$Actor_{002}$" stands for the external environment, "l_{002}" stands for an operation return formula, and "C_{002}" stands for a callee component as shown in Figure 6-2.

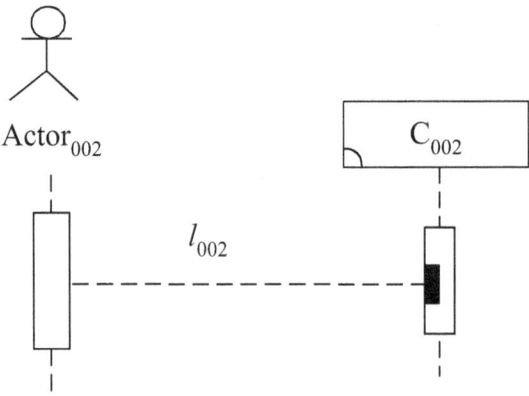

Figure 6-3　Prefix of n_{002}

As a third example, the n_{003} prefix defined as "**if** $cond_{003}$ **then** <OPERATION_CALL, $Actor_{003}$, l_{003}, C_{003}>" is a operation-based condition type_1 interaction under a certain (tautology excluded) condition in which "$cond_{003}$" stands for a (tautology excluded) condition, "OPERATION_CALL" stands for an operation call interaction tag, "$Actor_{003}$" stands for the external environment, "l_{003}" stands for an operation call formula, and "C_{003}" stands for a callee component as shown in Figure 6-4.

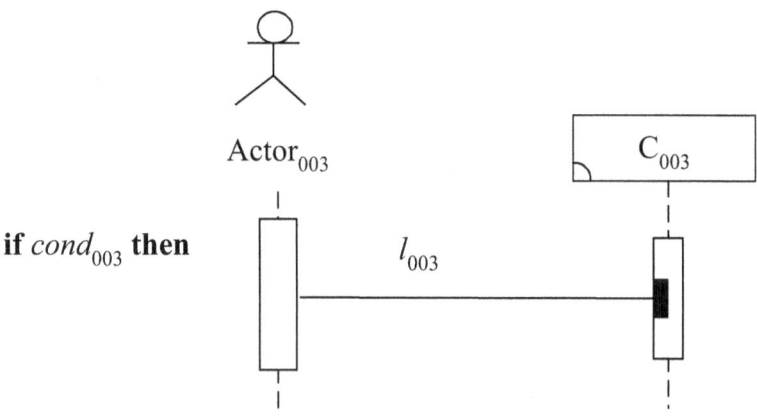

Figure 6-4　Prefix of n_{003}

As a fourth example, the n_{004} prefix defined as "**if** $cond_{004}$ **then** <OPERATION_RETURN, $Actor_{004}$, l_{004}, C_{004}>" is a operation-based condition type_1 interaction under a certain (tautology excluded) condition in which "$cond_{004}$" stands for a (tautology excluded) condition, "OPERATION_RETURN" stands for an operation return interaction tag, "$Actor_{004}$" stands for the external environment, "l_{004}" stands for an operation return formula, and "C_{004}" stands for a callee component as shown in Figure 6-5.

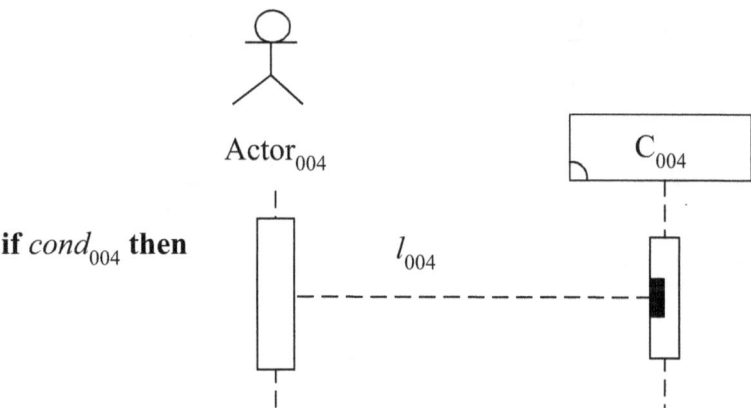

Figure 6-5　Prefix of n_{004}

As a fifth example, the z_{005} prefix defined as "<OPERATION_CALL, C_{005a}, l_{005}, C_{005b}>" or "**if** *true* **then** <OPERATION_CALL, C_{005a}, l_{005}, C_{005b}>" is an operation-based condition type_2 interaction under a tautology condition in which "OPERATION_CALL" stands for an operation call interaction tag, "C_{005a}" stands for a caller component, "l_{005}" stands for an operation call formula, and "C_{005b}" stands for a callee component as shown in Figure 6-6.

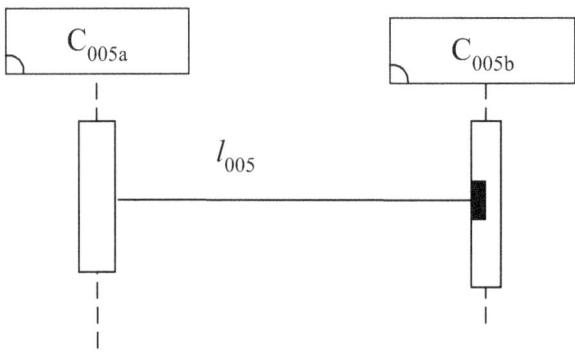

Figure 6-6　Prefix of z_{005}

As a sixth example, the z_{006} prefix defined as "<OPERATION_RETURN, C_{006a}, l_{006}, C_{006b}>" or "**if** *true* **then** <OPERATION_RETURN, C_{006a}, l_{006}, C_{006b}>" is an operation-based condition type_2 interaction under a tautology condition in which "OPERATION_RETURN" stands for an operation return interaction tag, "C_{006a}" stands for a caller component, "l_{006}" stands for an operation return formula, and "C_{006b}" stands for a callee component as shown in Figure 6-7.

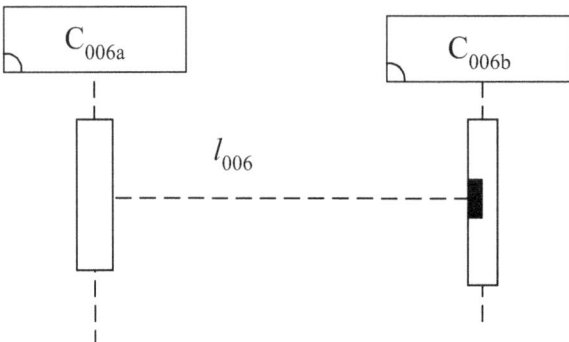

Figure 6-7　Prefix of z_{006}

As a seventh example, the z_{007} prefix defined as "**if** $cond_{007}$ **then** <OPERATION_CALL, C_{007a}, l_{007}, C_{007b}>" is a operation-based condition type_2 interaction under a certain (tautology excluded) condition in which "$cond_{007}$" stands for a (tautology excluded) condition, "OPERATION_CALL" stands for an operation call interaction tag, "C_{007a}" stands for a caller component, "l_{007}" stands for an operation call formula, and "C_{007b}" stands for a callee component as shown in Figure 6-8.

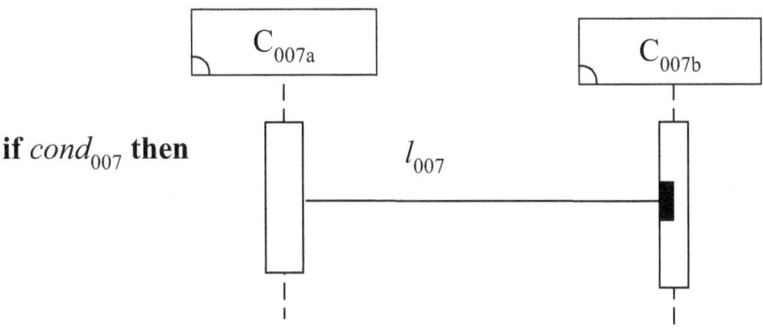

Figure 6-8   Prefix of z_{007}

As an eighth example, the z_{008} prefix defined as "**if** $cond_{008}$ **then** <OPERATION_RETURN, C_{008a}, l_{008}, C_{008b}>" is a operation-based condition type_2 interaction under a certain (tautology excluded) condition in which "$cond_{008}$" stands for a (tautology excluded) condition, "OPERATION_RETURN" stands for an operation return interaction tag, "C_{008a}" stands for a caller component, "l_{008}" stands for an operation return formula, and "C_{008b}" stands for a callee component as shown in Figure 6-9.

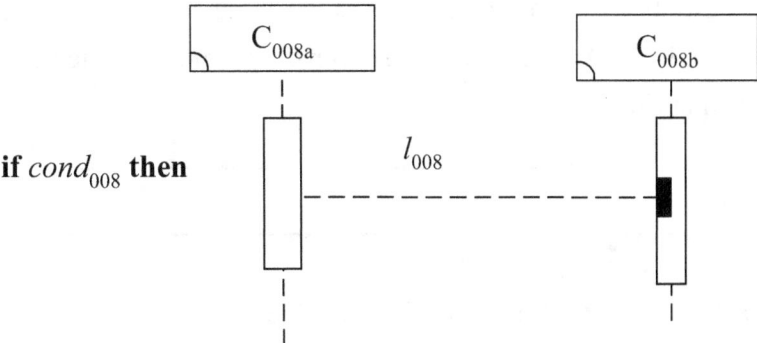

Figure 6-9   Prefix of z_{008}

6-3 Backus-Naur Form of Operation-Based Multi-Queue SBC Processes

The set of operation-based multi-queue SBC (i.e. Structure-Behavior Coalescence) processes is defined by the following BNF grammar, as shown in Figure 6-10.

(1) <System> ::= <FixIFD> {"∥" <FixIFD>}

(2) <FixIFD> ::= "fix(" <Process_Variable>"="<IFD>
 "●" <Process_Variable> ")"

(3) <IFD> ::= <Type_1_Expression> {"●" <Type_1_Or_2_Expression>}

(4) <Type_1_Or_2_Expression> ::= <Type_1_Expression>
 | <Type_2_Expression>

(5) <Type_1_Expression> ::= <Condition_Type_1_Interaction>
 {"+" <Condition_Type_1_Interaction>}

(6) <Type_2_Expression> ::= <Condition_Type_2_Interaction>
 {"+" <Condition_Type_2_Interaction>}

Figure 6-10　Backus-Naur Form
of Operation-Based Multi-Queue SBC Processes

6-3-1 Parallel Composition of One or More Recursive Interaction Flow Diagrams Defines the Operation-Based SBC Process of a System

Rule 1 describes that parallel composition of one or more recursive interaction flow diagrams (i.e. FixIFD) defines the operation-based multi-queue SBC process of a system, as shown in Figure 6-11.

Rule 1
<System> ::= <FixIFD> {"∥" <FixIFD>}

Figure 6-11　Rule 1

6-3-2 A Recursive Interaction Flow Diagram is the Recursion of an Interaction Flow Diagram

Rule 2 describes that a recursive interaction flow diagram (i.e. FixIFD) is defined by the recursion of an interaction flow diagram (i.e. IFD), as shown in Figure 6-12.

Rule 2
<FixIFD> ::= **fix**(" <Process_Variable> "=" <IFD> " • " <Process_Variable> ")"

Figure 6-12   Rule 2

6-3-3 An Interaction Flow Diagram is a Type_1 Expression Followed by Zero or More Type_1_Or_Type_2 Expressions

Rule 3 describes that an interaction flow diagram (i.e. IFD) is defined by a type_1 expression (i.e. Type_1_Expression) followed by zero or more type_1_or_type_2 expressions (i.e. Type_1_Or_2_Expression), as shown in Figure 6-13.

Rule 3
<IFD> ::= <Type_1_Expression> {" • " <Type_1_Or_2_Expression>}

Figure 6-13   Rule 3

6-3-4 Type_1_Or_2 Expression is either Type_1 or Type_2

Rule 4 describes that the type_1_or_2 expression (i.e. Type_1_Or_2_Expression) is either a type_1 expression (i.e. Type_1_Expression) or a type_2 expression (i.e. Type_2_Expression), as shown in Figure 6-14.

Rule 4
<Type_1_Or_2_Expression> ::= <Type_1_Expression> 　　　　　　　　　　　　| <Type_2_Expression>

Figure 6-14　　Rule 4

6-3-5 Summation of One or More Condition Type_1 Interactions Defines the Type_1 Expression

Rule 5 describes that the summation of one or more condition type_1 interactions (i.e. Condition_Type_1_Interaction) defines the type_1 expression (i.e. Type_1_Expression), as shown in Figure 6-15.

Rule 5
<Type_1_Expression> ::= <Condition_Type_1_Interaction> 　　　　　　　　　　　{"+" <Condition_Type_1_Interaction>}

Figure 6-15　　Rule 5

6-3-6 Summation of One or More Condition Type_2 Interactions Defines the Type_2 Expression

Rule 6 describes that the summation of one or more condition type_2 interactions (i.e. Condition_Type_2_Interaction) defines the type_2 expression (i.e. Type_1_Expression), as shown in Figure 6-16.

Rule 6
<Type_2_Expression> ::= <Condition_Type_2_Interaction> {"+" <Condition_Type_2_Interaction>}

Figure 6-16   Rule 6

6-4 Examples of Type_1 Expression

As a first example, consider the type-1 expression E_{101} is defined as "n_{101}" and the n_{101} prefix is defined as "<OPERATION_CALL, $Actor_{101}$, l_{101}, C_{101}>". The Backus-Naur Form tree of E_{101} is shown in Figure 6-17.

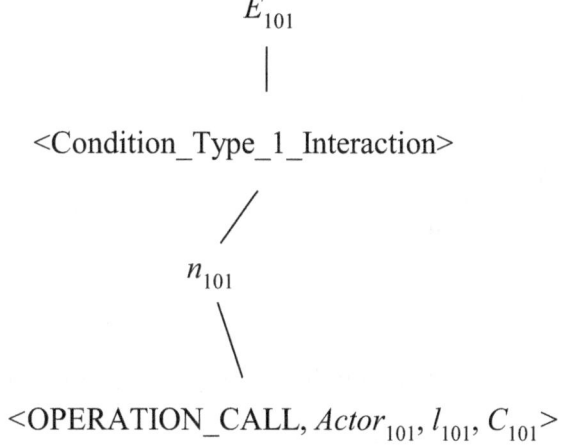

Figure 6-17   Backus-Naur Form Tree of the E_{101} Type-1 Expression

As a second example, consider the type-1 expression E_{102} is defined as "n_{102}" and the n_{102} prefix is defined as "<OPERATION_RETURN, $Actor_{102}$, l_{102}, C_{102}>". The Backus-Naur Form tree of E_{102} is shown in Figure 6-18.

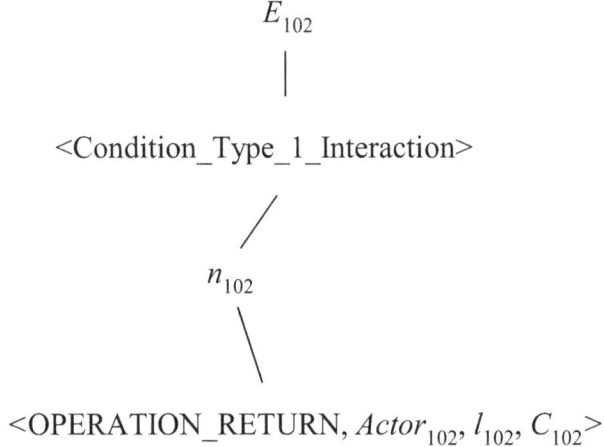

Figure 6-18   Backus-Naur Form Tree of the E_{102} Type-1 Expression

As a third example, consider the type-1 expression E_{103} is defined as "n_{103}" and the n_{103} prefix is defined as "**if** $cond_{103}$ **then** "<OPERATION_CALL, $Actor_{103}$, l_{103}, C_{103}>". The Backus-Naur Form tree of E_{103} is shown in Figure 6-19.

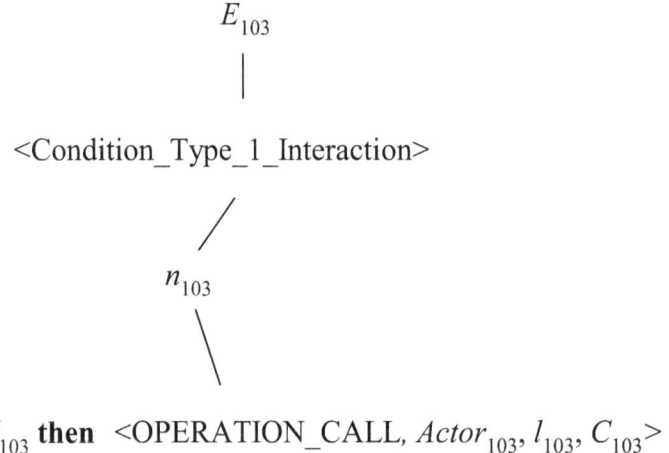

Figure 6-19   Backus-Naur Form Tree of the E_{103} Type-1 Expression

As a fourth example, consider the type-1 expression E_{104} is defined as "n_{104}" and the n_{104} prefix is defined as "**if** $cond_{104}$ **then** "<OPERATION_RETURN, $Actor_{104}$, l_{104}, C_{104}>". The Backus-Naur Form tree of E_{104} is shown in Figure 6-20.

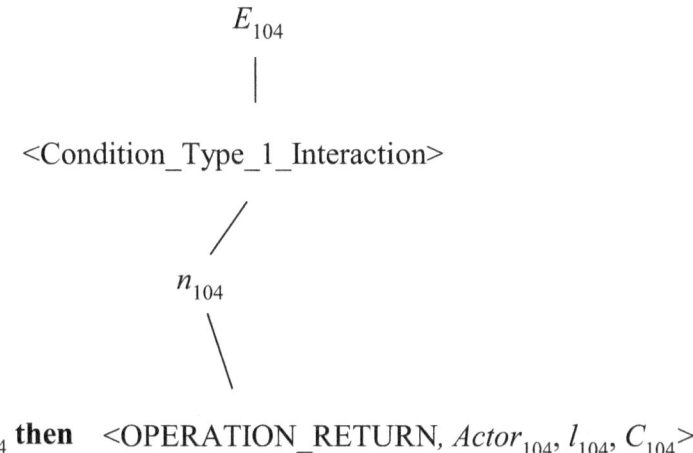

Figure 6-20   Backus-Naur Form Tree of the E_{104} Type-1 Expression

As a fifth example, consider the type-1 expression E_{105} is defined as "$n_{105a}+n_{105b}$" and the n_{105a}, n_{105b} prefixes are defined respectively as "<OPERATION_CALL, $Actor_{105a}$, l_{105a}, C_{105a}>", "<OPERATION_CALL, $Actor_{105b}$, l_{105b}, C_{105b}>". The Backus-Naur Form tree of E_{105} is shown in Figure 6-21.

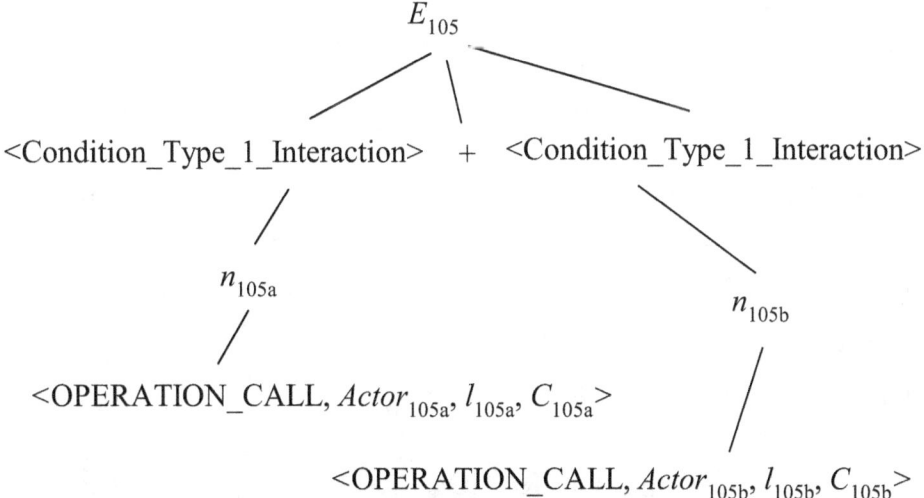

Figure 6-21   Backus-Naur Form Tree of the E_{105} Type-1 Expression

As a sixth example, consider the type-1 expression E_{106} is defined as "$n_{106a}+n_{106b}$" and the n_{106a}, n_{106b} prefixes are defined respectively as "<OPERATION_RETURN, $Actor_{106a}$, l_{106a}, C_{106a}>", "<OPERATION_RETURN, $Actor_{106b}$, l_{106b}, C_{106b}>". The Backus-Naur Form tree of E_{106} is shown in Figure 6-22.

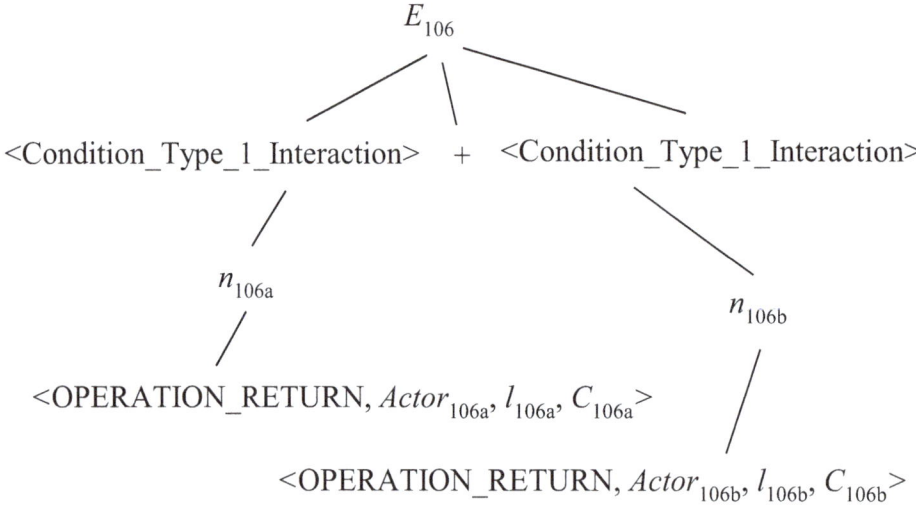

Figure 6-22   Backus-Naur Form Tree of the E_{106} Type-1 Expression

As a seventh example, consider the type-1 expression E_{107} is defined as "$n_{107a}+n_{107b}$" and the n_{107a}, n_{107b} prefixes are defined respectively as "<OPERATION_CALL, $Actor_{107a}$, l_{107a}, C_{107a}>", "**if** $cond_{107b}$ **then** <OPERATION_CALL, $Actor_{107b}$, l_{107b}, C_{107b}>". The Backus-Naur Form tree of E_{107} is shown in Figure 6-23.

Figure 6-23   Backus-Naur Form Tree of the E_{107} Type-1 Expression

As an eighth example, consider the type-1 expression E_{108} is defined as "$n_{108a}+n_{108b}$" and the n_{108a}, n_{108b} prefixes are defined respectively as "<OPERATION_RETURN, $Actor_{108a}$, l_{108a}, C_{108a}>", "**if** $cond_{108b}$ **then** <OPERATION_RETURN, $Actor_{108b}$, l_{108b}, C_{108b}>". The Backus-Naur Form tree of E_{108} is shown in Figure 6-24.

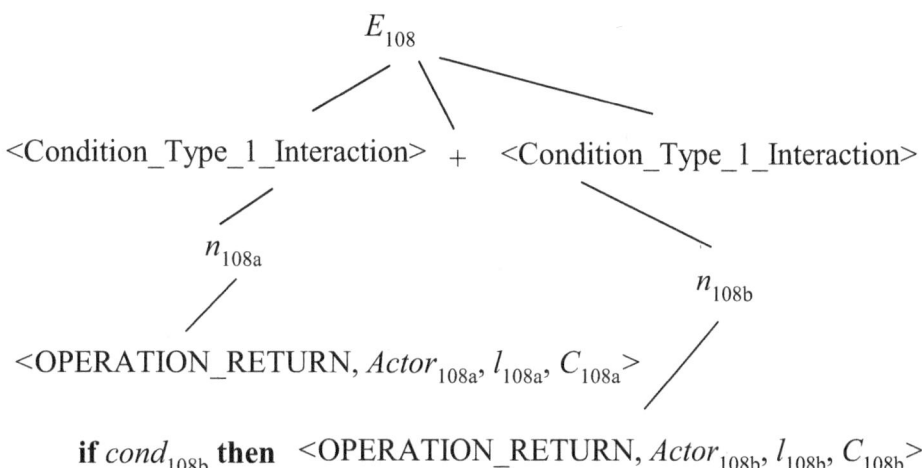

Figure 6-24   Backus-Naur Form Tree of the E_{108} Type-1 Expression

As a ninth example, consider the type-1 expression E_{109} is defined as "$n_{109a}+n_{109b}$" and the n_{109a}, n_{109b} prefixes are defined respectively as "**if** $cond_{109a}$ **then** "<OPERATION_CALL, $Actor_{109a}$, l_{109a}, C_{109a}>", "**if** $cond_{109b}$ **then** <OPERATION_CALL, $Actor_{109b}$, l_{109b}, C_{109b}>". The Backus-Naur Form tree of E_{109} is shown in Figure 6-25.

Figure 6-25   Backus-Naur Form Tree of the E_{109} Type-1 Expression

As a tenth example, consider the type-1 expression E_{110} is defined as "$n_{110a}+n_{110b}$" and the n_{110a}, n_{110b} prefixes are defined respectively as "**if** $cond_{110a}$ **then** <OPERATION_RETURN, $Actor_{110a}$, l_{110a}, C_{110a}>", "**if** $cond_{110b}$ **then** <OPERATION_RETURN, $Actor_{110b}$, l_{110b}, C_{110b}>". The Backus-Naur Form tree of E_{110} is shown in Figure 6-26.

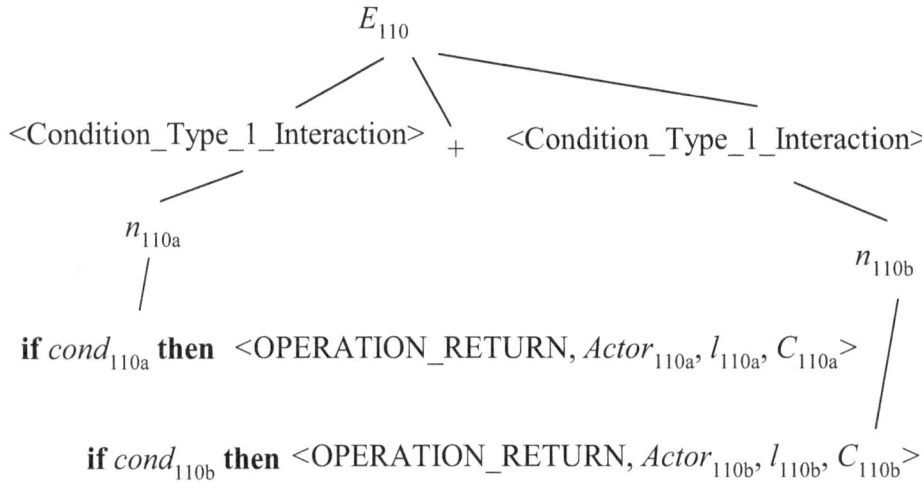

Figure 6-26 Backus-Naur Form Tree of the E_{110} Type-1 Expression

As an eleventh example, consider the type-1 expression E_{111} is defined as "$n_{111a}+n_{111b}$" and the n_{111a}, n_{111b} prefixes are defined respectively as "<OPERATION_CALL, $Actor_{111}$, l_{111a}, C_{111a}>", "<OPERATION_CALL, $Actor_{111}$, l_{111b}, C_{111b}>". The Backus-Naur Form tree of E_{111} is shown in Figure 6-27.

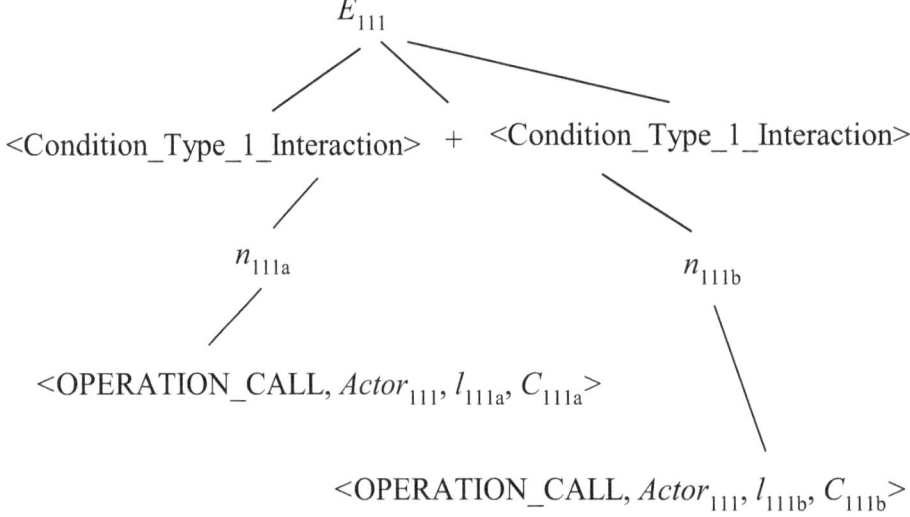

Figure 6-27 Backus-Naur Form Tree of the E_{111} Type-1 Expression

As a twelfth example, consider the type-1 expression E_{112} is defined as "$n_{112a}+n_{112b}$" and the n_{112a}, n_{112b} prefixes are defined respectively as "<OPERATION_RETURN, $Actor_{112}$, l_{112a}, C_{112a}>", "<OPERATION_RETURN, $Actor_{112}$, l_{112b}, C_{112b}>". The Backus-Naur Form tree of E_{112} is shown in Figure 6-28.

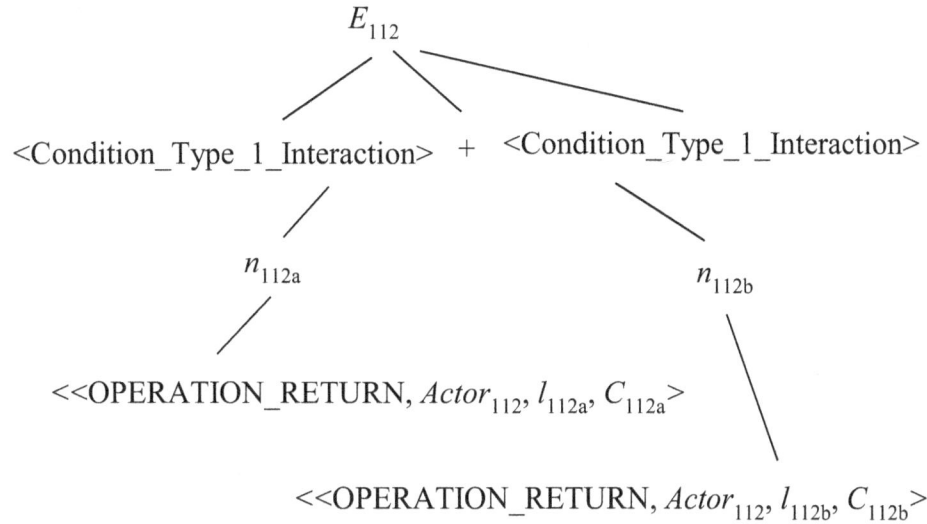

Figure 6-28   Backus-Naur Form Tree of the E_{112} Type-1 Expression

As a thirteenth example, consider the type-1 expression E_{113} is defined as "$n_{113a}+n_{113b}$" and the n_{113a}, n_{113b} prefixes are defined respectively as "<OPERATION_CALL, $Actor_{113}$, l_{113a}, C_{113a}>", "**if** $cond_{113b}$ **then** <OPERATION_CALL, $Actor_{113}$, l_{113b}, C_{113b}>". The Backus-Naur Form tree of E_{113} is shown in Figure 6-29.

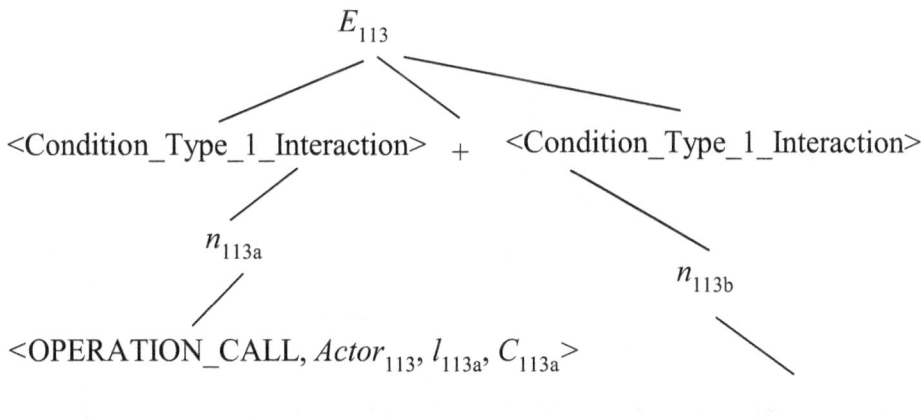

Figure 6-29   Backus-Naur Form Tree of the E_{113} Type-1 Expression

As a fourteenth example, consider the type-1 expression E_{114} is defined as "$n_{114a}+n_{114b}$" and the n_{114a}, n_{114b} prefixes are defined respectively as "<OPERATION_RETURN, $Actor_{114}$, l_{114a}, C_{114a}>", "**if** $cond_{114b}$ **then** <OPERATION_RETURN, $Actor_{114}$, l_{114b}, C_{114b}>". The Backus-Naur Form tree of E_{114} is shown in Figure 6-30.

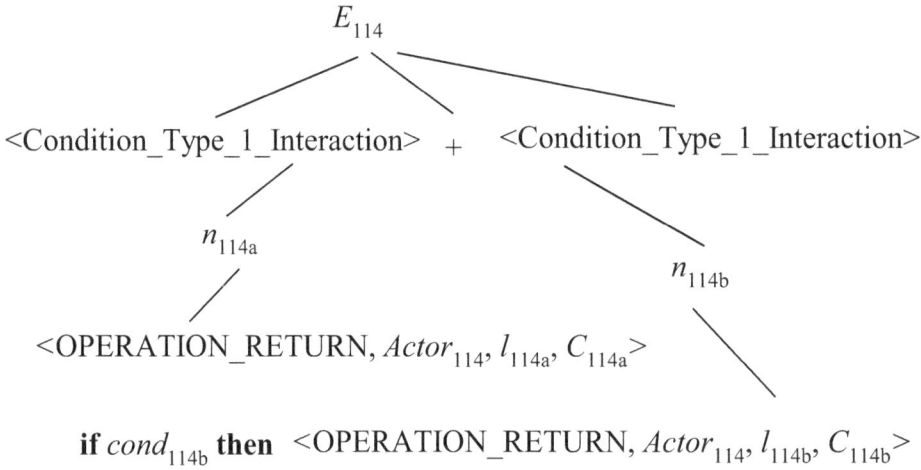

Figure 6-30   Backus-Naur Form Tree of the E_{114} Type-1 Expression

As a fifteenth example, consider the type-1 expression E_{115} is defined as "$n_{115a}+n_{115b}$" and the n_{115a}, n_{115b} prefixes are defined respectively as "**if** $cond_{115a}$ **then** <OPERATION_CALL, $Actor_{115a}$, l_{115a}, C_{115a}>", "**if** $cond_{115b}$ **then** <OPERATION_CALL, $Actor_{115b}$, l_{115b}, C_{115b}>". The Backus-Naur Form tree of E_{115} is shown in Figure 6-31.

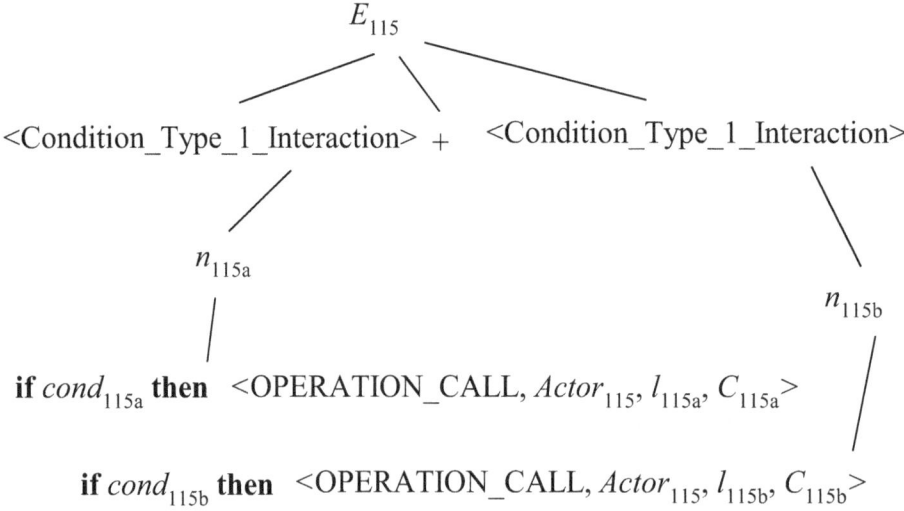

Figure 6-31   Backus-Naur Form Tree of the E_{115} Type-1 Expression

As a sixteenth example, consider the type-1 expression E_{116} is defined as "$n_{116a}+n_{116b}$" and the n_{116a}, n_{116b} prefixes are defined respectively as "**if** $cond_{116a}$ **then** <OPERATION_RETURN, $Actor_{116a}$, l_{116a}, C_{116a}>", "**if** $cond_{116b}$ **then** <OPERATION_RETURN, $Actor_{116b}$, l_{116b}, C_{116b}>". The Backus-Naur Form tree of E_{116} is shown in Figure 6-32.

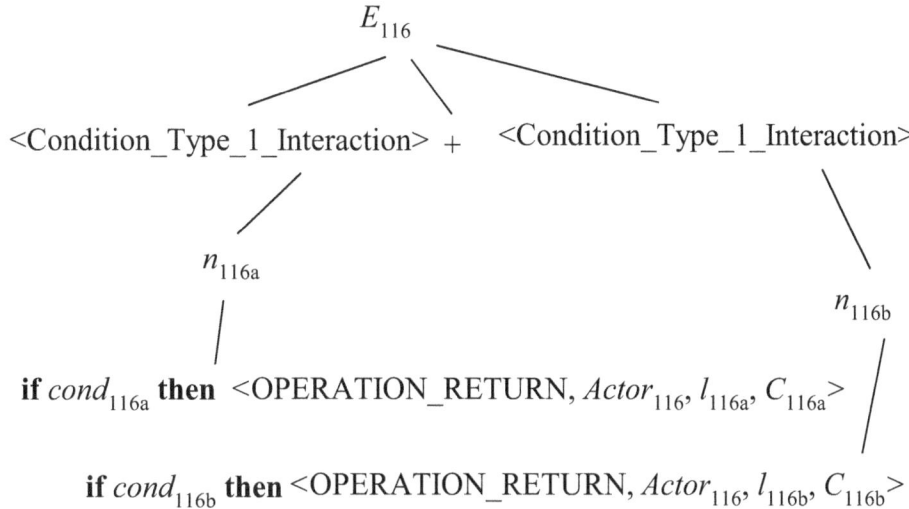

Figure 6-32   Backus-Naur Form Tree of the E_{116} Type-1 Expression

6-5 Examples of Type_2 Expression

As a first example, consider the type-2 expression E_{201} is defined as "z_{201}" and the z_{201} prefix is defined as "<OPERATION_CALL, C_{201a}, l_{201}, C_{201b}>". The Backus-Naur Form tree of E_{201} is shown in Figure 6-33.

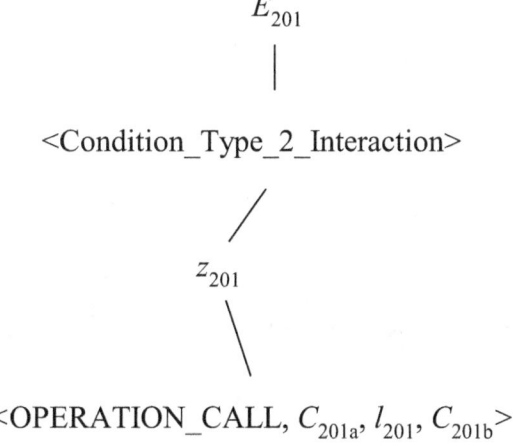

Figure 6-33   Backus-Naur Form Tree of the E_{201} Type-2 Expression

As a second example, consider the type-2 expression E_{202} is defined as "z_{202}" and the z_{202} prefix is defined as "<OPERATION_RETURN, C_{202a}, l_{202}, C_{202b}>". The Backus-Naur Form tree of E_{202} is shown in Figure 6-34.

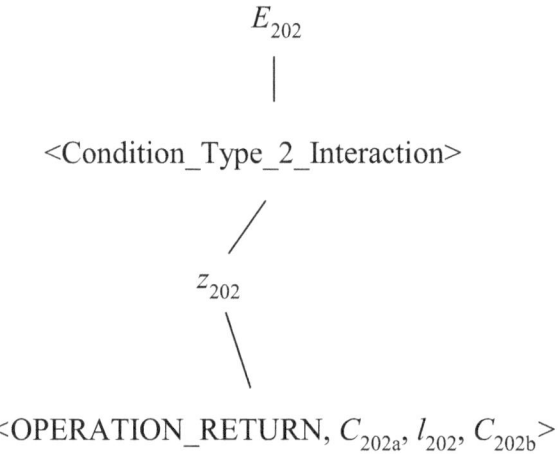

Figure 6-34   Backus-Naur Form Tree of the E_{202} Type-2 Expression

As a third example, consider the type-2 expression E_{203} is defined as "z_{203}" and the z_{203} prefix is defined as "**if** $cond_{203}$ **then** <OPERATION_CALL, C_{203a}, l_{203}, C_{203b}>". The Backus-Naur Form tree of E_{203} is shown in Figure 6-35.

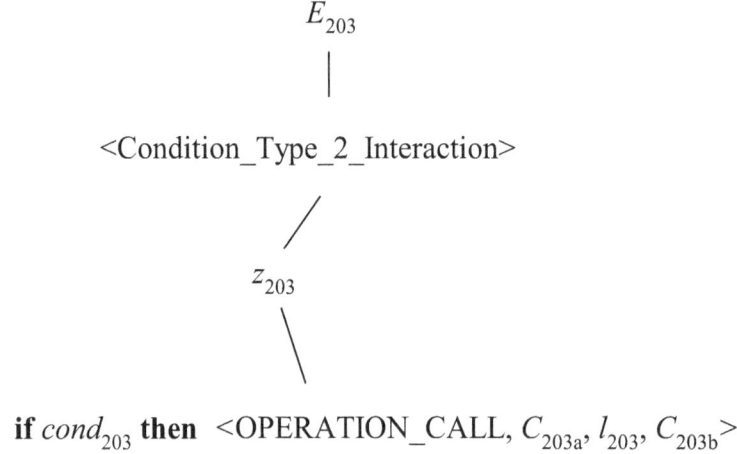

Figure 6-35   Backus-Naur Form Tree of the E_{203} Type-2 Expression

As a fourth example, consider the type-2 expression E_{204} is defined as "z_{204}" and the z_{204} prefix is defined as "**if** $cond_{204}$ **then** <OPERATION_RETURN, C_{204a}, l_{204}, C_{204b}>". The Backus-Naur Form tree of E_{204} is shown in Figure 6-36.

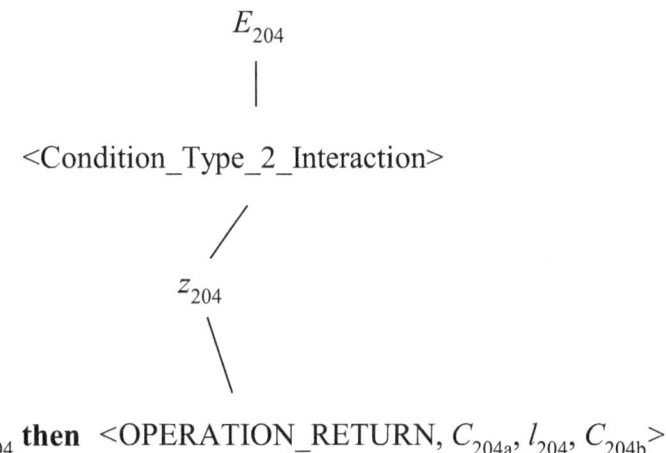

Figure 6-36   Backus-Naur Form Tree of the E_{204} Type-2 Expression

As a fifth example, consider the type-2 expression E_{205} is defined as "$z_{205a}+z_{205b}$" and the z_{205a}, z_{205b} prefixes are defined respectively as "<OPERATION_CALL, C_{205a}, l_{205a}, C_{205c}>", "<OPERATION_CALL, C_{205b}, l_{205b}, C_{205d}>". The Backus-Naur Form tree of E_{205} is shown in Figure 6-37.

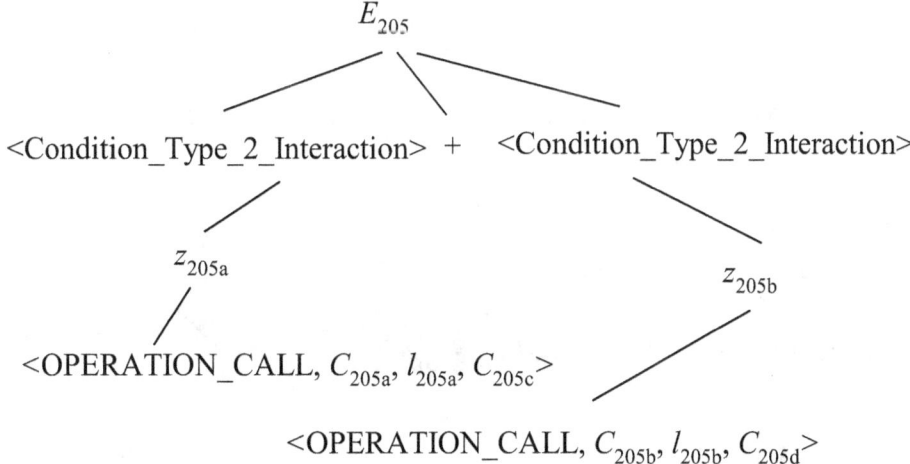

Figure 6-37   Backus-Naur Form Tree of the E_{205} Type-2 Expression

As a sixth example, consider the type-2 expression E_{206} is defined as "$z_{206a}+z_{206b}$" and the z_{206a}, z_{206b} prefixes are defined respectively as "<OPERATION_RETURN, C_{206a}, l_{206a}, C_{206c}>", "<OPERATION_RETURN, C_{206b}, l_{206b}, C_{206d}>". The Backus-Naur Form tree of E_{206} is shown in Figure 6-38.

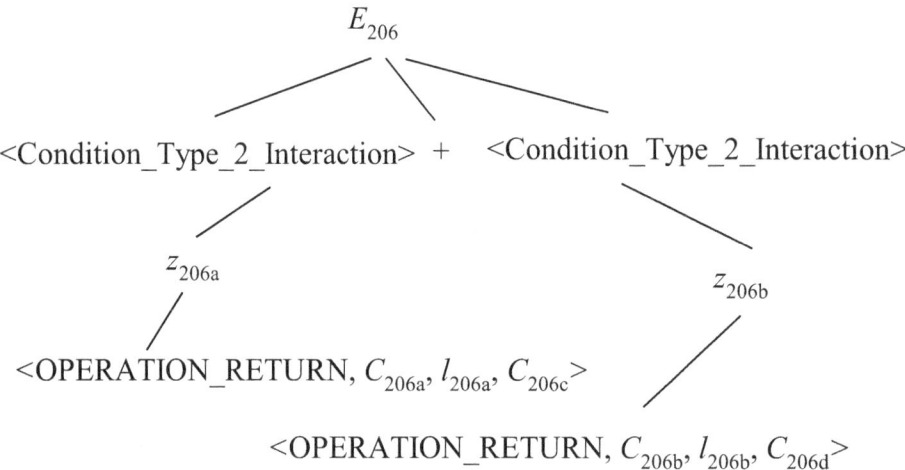

Figure 6-38   Backus-Naur Form Tree of the E_{206} Type-2 Expression

As a seventh example, consider the type-2 expression E_{207} is defined as "$z_{207a}+z_{207b}$" and the z_{207a}, z_{207b} prefixes are defined respectively as "<OPERATION_CALL, C_{207a}, l_{207a}, C_{207c}>", "**if** $cond_{207b}$ **then** <OPERATION_CALL, C_{207b}, l_{207b}, C_{207d}>". The Backus-Naur Form tree of E_{207} is shown in Figure 6-39.

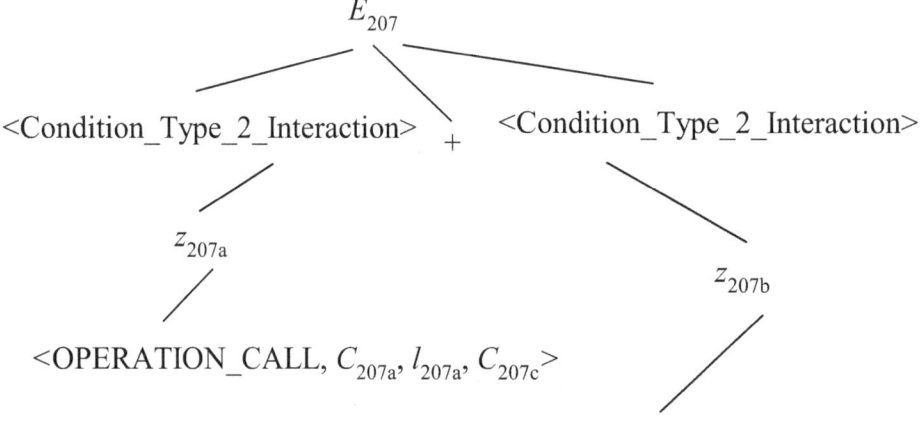

Figure 6-39   Backus-Naur Form Tree of the E_{207} Type-2 Expression

As an eighth example, consider the type-2 expression E_{208} is defined as "$z_{208a}+z_{208b}$" and the z_{208a}, z_{208b} prefixes are defined respectively as "<OPERATION_RETURN, C_{208a}, l_{208a}, C_{208c}>", "**if** $cond_{208b}$ **then** <OPERATION_RETURN, C_{208b}, l_{208b}, C_{208d}>". The Backus-Naur Form tree of E_{208} is shown in Figure 6-40.

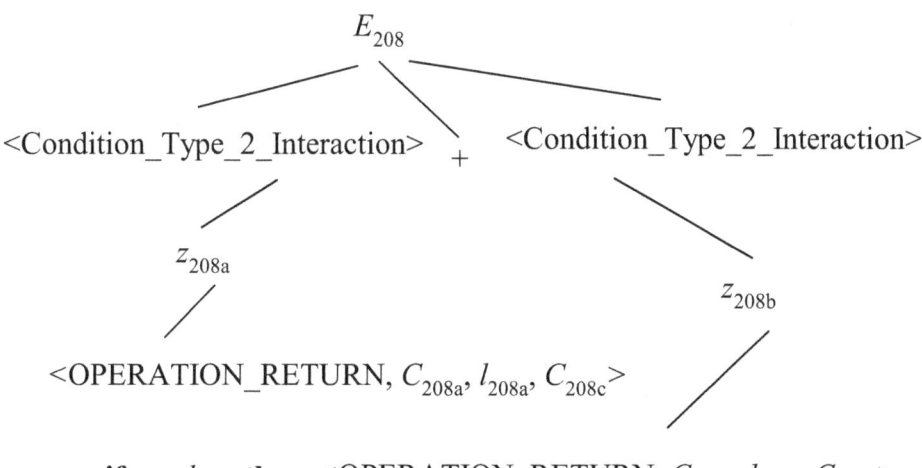

Figure 6-40   Backus-Naur Form Tree of the E_{208} Type-2 Expression

As a ninth example, consider the type-2 expression E_{209} is defined as "$z_{209a}+z_{209b}$" and the z_{209a}, z_{209b} prefixes are defined respectively as "**if** $cond_{209a}$ **then** <OPERATION_CALL, C_{209a}, l_{209a}, C_{209c}>", "**if** $cond_{209b}$ **then** <OPERATION_CALL, C_{209b}, l_{209b}, C_{209d}>". The Backus-Naur Form tree of E_{209} is shown in Figure 6-41.

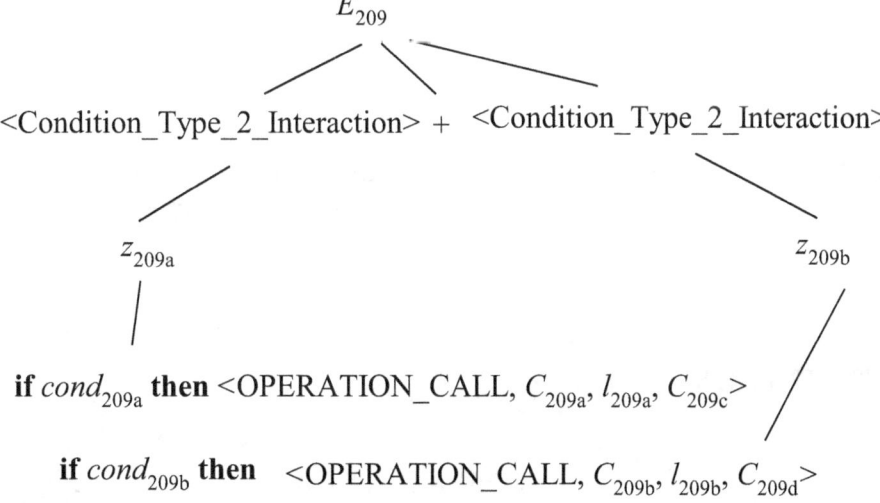

Figure 6-41   Backus-Naur Form Tree of the E_{209} Type-2 Expression

As a tenth example, consider the type-2 expression E_{210} is defined as "$z_{210a}+z_{210b}$" and the z_{210a}, z_{210b} prefixes are defined respectively as "**if** $cond_{210a}$ **then** <OPERATION_RETURN, C_{210a}, l_{210a}, C_{210c}>", "**if** $cond_{210b}$ **then** <OPERATION_RETURN, C_{210b}, l_{210b}, C_{210d}>". The Backus-Naur Form tree of E_{210} is shown in Figure 6-42.

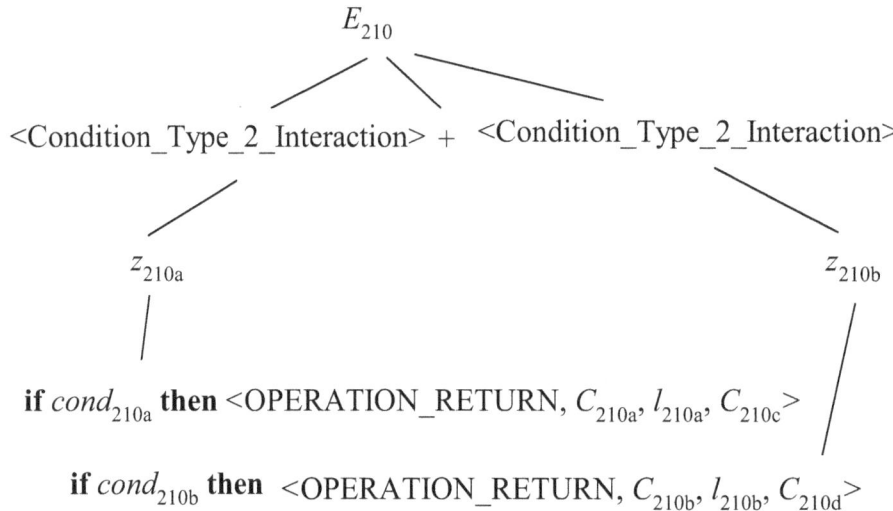

Figure 6-42   Backus-Naur Form Tree of the E_{210} Type-2 Expression

As an eleventh example, consider the type-2 expression E_{211} is defined as "$z_{211a}+z_{211b}$" and the z_{211a}, z_{211b} prefixes are defined respectively as "<OPERATION_CALL, C_{211}, l_{211a}, C_{211a}>", "<OPERATION_CALL, C_{211}, l_{211b}, C_{211b}>". The Backus-Naur Form tree of E_{211} is shown in Figure 6-43.

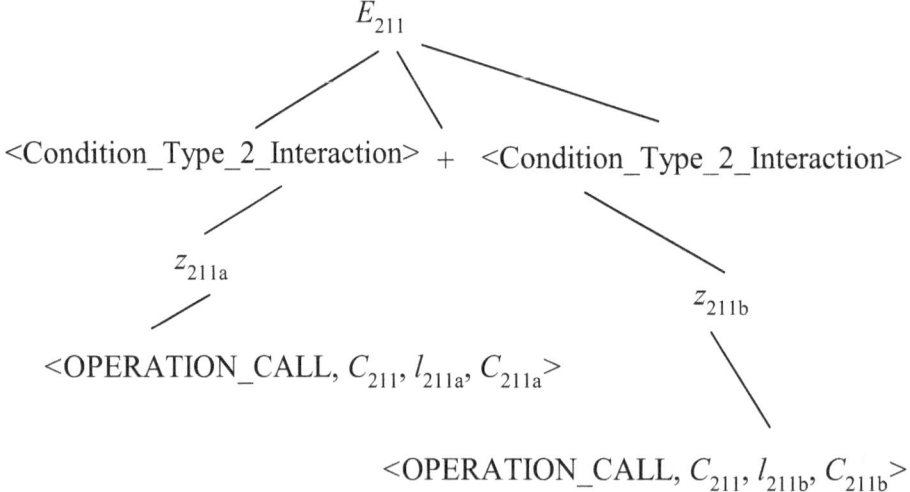

Figure 6-43   Backus-Naur Form Tree of the E_{211} Type-2 Expression

As a twelfth example, consider the type-2 expression E_{212} is defined as "$z_{212a}+z_{212b}$" and the z_{212a}, z_{212b} prefixes are defined respectively as "<OPERATION_RETURN, C_{212}, l_{212a}, C_{212a}>", "<OPERATION_RETURN, C_{212}, l_{212b}, C_{212b}>". The Backus-Naur Form tree of E_{212} is shown in Figure 6-44.

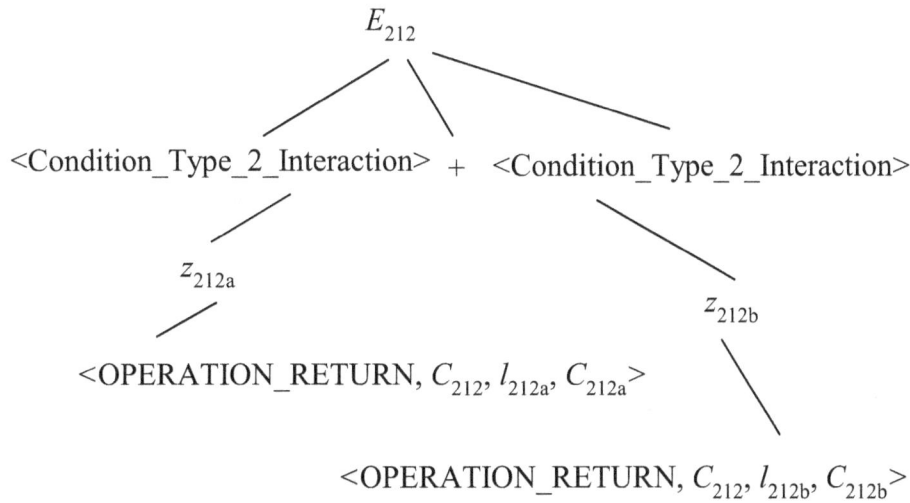

Figure 6-44   Backus-Naur Form Tree of the E_{212} Type-2 Expression

As a thirteenth example, consider the type-2 expression E_{213} is defined as "$z_{213a}+z_{213b}$" and the z_{213a}, z_{213b} prefixes are defined respectively as "<OPERATION_CALL, C_{213}, l_{213a}, C_{213a}>", "**if** $cond_{213b}$ **then** <OPERATION_CALL, C_{213}, l_{213b}, C_{213b}>". The Backus-Naur Form tree of E_{213} is shown in Figure 6-45.

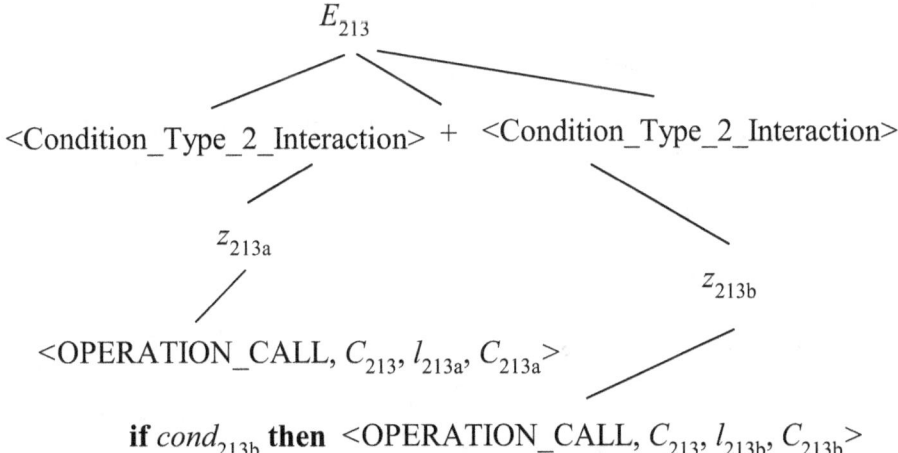

Figure 6-45   Backus-Naur Form Tree of the E_{213} Type-2 Expression

As a fourteenth example, consider the type-2 expression E_{214} is defined as "$z_{214a}+z_{214b}$" and the z_{214a}, z_{214b} prefixes are defined respectively as "<OPERATION_RETURN, C_{214}, l_{214a}, C_{214a}>", "**if** $cond_{214b}$ **then** <OPERATION_RETURN, C_{214}, l_{214b}, C_{214b}>". The Backus-Naur Form tree of E_{214} is shown in Figure 6-46.

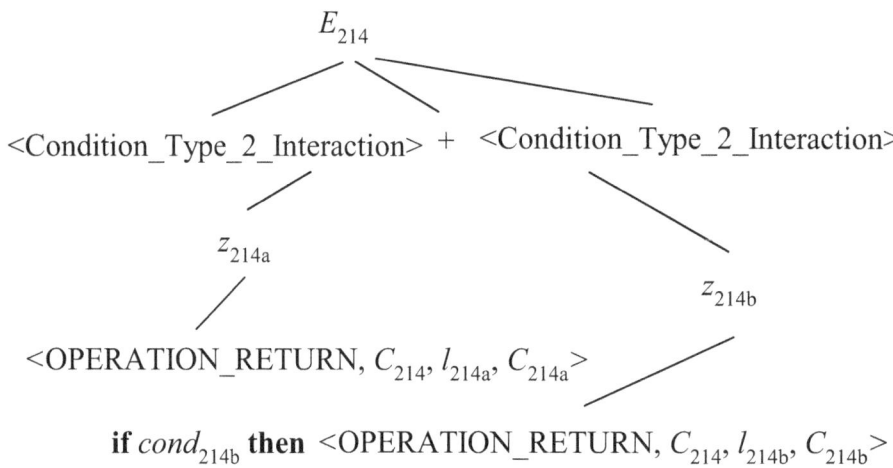

Figure 6-46   Backus-Naur Form Tree of the E_{214} Type-2 Expression

As a fifteenth example, consider the type-2 expression E_{215} is defined as "$z_{215a}+z_{215b}$" and the z_{215a}, z_{215b} prefixes are defined respectively as "**if** $cond_{215a}$ **then** <OPERATION_CALL, C_{215}, l_{215a}, C_{215a}>", "**if** $cond_{215b}$ **then** <OPERATION_CALL, C_{215}, l_{215b}, C_{215b}>". The Backus-Naur Form tree of E_{215} is shown in Figure 6-47.

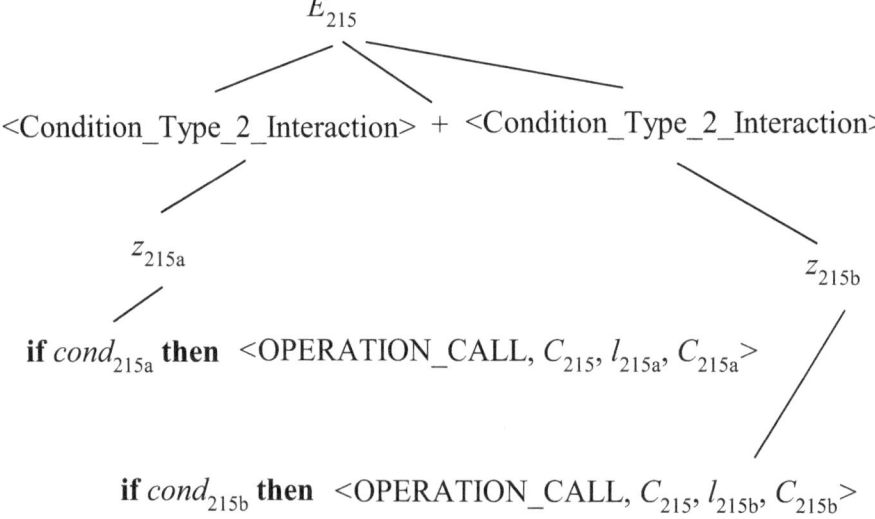

Figure 6-47   Backus-Naur Form Tree of the E_{215} Type-2 Expression

As a sixteenth example, consider the type-2 expression E_{216} is defined as "$z_{216a}+z_{216b}$" and the z_{216a}, z_{216b} prefixes are defined respectively as "**if** $cond_{216a}$ **then** <OPERATION_RETURN, C_{216}, l_{216a}, C_{216a}>", "**if** $cond_{216b}$ **then** <OPERATION_RETURN, C_{216}, l_{216b}, C_{216b}>". The Backus-Naur Form tree of E_{216} is shown in Figure 6-48.

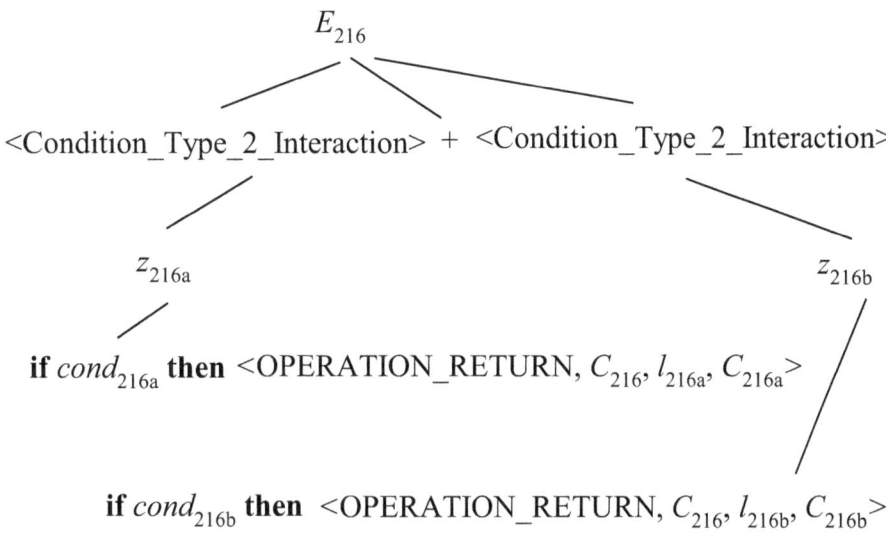

Figure 6-48　Backus-Naur Form Tree of the E_{216} Type-2 Expression

Chapter 7: Transitional Semantics of Operation-Based Multi-Queue SBC Process Algebra

In the chapter, we illustrate in detail those operation-based multi-queue SBC process algebra transitional semantics which regards the system definition of a system.

7-1 Transitional Semantics

In giving meaning to the operation-based multi-queue SBC process algebra, we shall use the following transition system

$$(\Psi, T, TR)$$

which consists of a set Ψ of process expressions, a set T of "condition type_1 or type_2 interaction", and a transition relation $TR \subseteq \Psi \times T \times \Psi$ where $(E_i, t, E_j) \in T$ is denoted by $E_i \xrightarrow{t} E_j$.

The semantics for Ψ consists in the transition rules of each transition relation TR over $\Psi \times T \times \Psi$. These transition rules will follow the structure of expressions.

As shown in Figure 7-1, we give the complete set of transition rules; the names Prefix, Sum, Parallel, Recursion, and Constant indicate that the rules are associated respectively with Prefix, Summation, Parallel Composition, and Recursion and with Constants.

$$
\text{Prefix} \qquad \frac{}{t \bullet E \xrightarrow{t} E}
$$

$$
\text{Sum}_j \qquad \frac{E_j \xrightarrow{t} E'_j}{\sum_{i \in I} E_i \xrightarrow{t} E'_j} (j \in I)
$$

$$
\text{Parallel}_1 \qquad \frac{E \xrightarrow{t} E'}{E \parallel F \xrightarrow{t} E' \parallel F}
$$

$$
\text{Parallel}_2 \qquad \frac{F \xrightarrow{t} F'}{E \parallel F \xrightarrow{t} E \parallel F'}
$$

$$
\text{Recursion} \qquad \frac{E\{\mathbf{fix}(X=E)/X\} \xrightarrow{t} E'}{\mathbf{fix}(X=E) \xrightarrow{t} E'}
$$

$$
\text{Constant} \qquad \frac{P \xrightarrow{t} P'}{A \xrightarrow{t} P'} (A \stackrel{\text{def}}{=} P)
$$

Figure 7-1   Transition Rules for the
Operation-Based Multi-Queue SBC Process Algebra

We said that the set of rules shown in Figure 7-1 is complete; by this we mean that there are no transitions except those which can be inferred by the rules.

7-2 Rule of Prefix

The rule for Prefix, shown in Figure 7-2, can be read as follows: Under any circumstances, we always infer $t \bullet E \xrightarrow{t} E$. That is, an expression, with a "condition type_1 or type_2 interaction" prefixed to it, will use this "condition type_1 or type_2 interaction" to accomplish the transition.

$$t \bullet E \xrightarrow{t} E$$

Figure 7-2   Rule of Prefix

7-3 Rule of Summation

The rule for Summation, shown in Figure 7-3, can be read as follows: If any one summand E_j of the sum $\sum_{i \in I} E_i$ has a "condition type_1 or type_2 interaction", then the whole sum also has that "condition type_1 or type_2 interaction".

$$\frac{E_j \xrightarrow{t} E'_j}{\sum_{i \in I} E_i \xrightarrow{t} E'_j} (j \in I)$$

Figure 7-3   Rule of Summation

7-4 Rule of Parallel Composition

There are two transition rules for parallel composition. Rule Parallel$_1$, as shown in Figure 7-4, indicates that from $E \xrightarrow{t} E'$ we shall infer $E\|F \xrightarrow{t} E'\|F$.

$$E \xrightarrow{t} E'$$
$$\overline{E \parallel F \xrightarrow{t} E' \parallel F}$$

Figure 7-4   Rule Parallel$_1$

Rule Parallel$_2$, as shown in Figure 7-5, indicates that from $F \xrightarrow{t} F'$ we shall infer $E \parallel F \xrightarrow{t} E \parallel F'$.

$$F \xrightarrow{t} F'$$
$$\overline{E \parallel F \xrightarrow{t} E \parallel F'}$$

Figure 7-5   Rule Parallel$_2$

7-5 Rule of Recursion

The rule for Recursion, shown in Figure 7-6, can be read as follows: This says that any "condition type_1 or type_2 interaction" which may be inferred for the **fix** expression 'unwound' once (by substituting itself for its bound variable) may be inferred for the **fix** expression itself.

$$E\{\mathbf{fix}(X{=}E)/X\} \xrightarrow{t} E'$$
$$\overline{\mathbf{fix}(X{=}E) \xrightarrow{t} E'}$$

Figure 7-6   Rule of Recursion

7-6 Rule of Constants

The rule for Constants, shown in Figure 7-7, can be read as follows: the rule of Constants asserts that each Constant has the same transitions as its defining expression.

$$\frac{P \xrightarrow{t} P'}{A \xrightarrow{t} P'} \quad (A \stackrel{\text{def}}{=} P)$$

Figure 7-7　Rule of Constants

PART III: CASES STUDY

Chapter 8: Operation-Based Multi-Queue SBC Process of the Restaurant

In this chapter, we use the operation-based multi-queue SBC process algebra to define the *Restaurant* as shown in Figure 8-1.

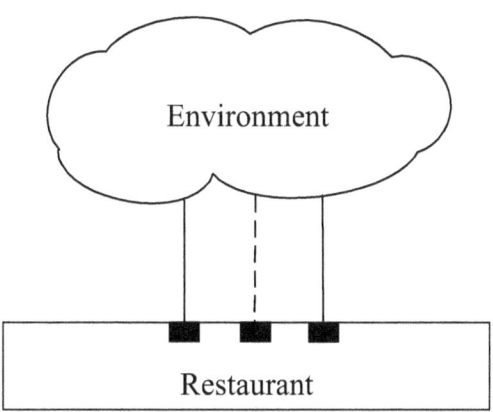

Figure 8-1   Systems Definition of the *Restaurant*

8-1 BNF Tree of the Restaurant

The operation-based multi-queue SBC process of the *Restaurant*, A_{501}, is defined as "$\mathbf{fix}(X_{501}=n_{501}\bullet(z_{502}+z_{503}+z_{504})\bullet(z_{505}+z_{506}+z_{507})\bullet n_{508}\bullet X_{501})\|\mathbf{fix}(X_{502}=n_{509}\bullet X_{502})$".

We draw the operation-based multi-queue SBC process algebra Backus-Naur Form tree of A_{501} as shown in Figure 8-2.

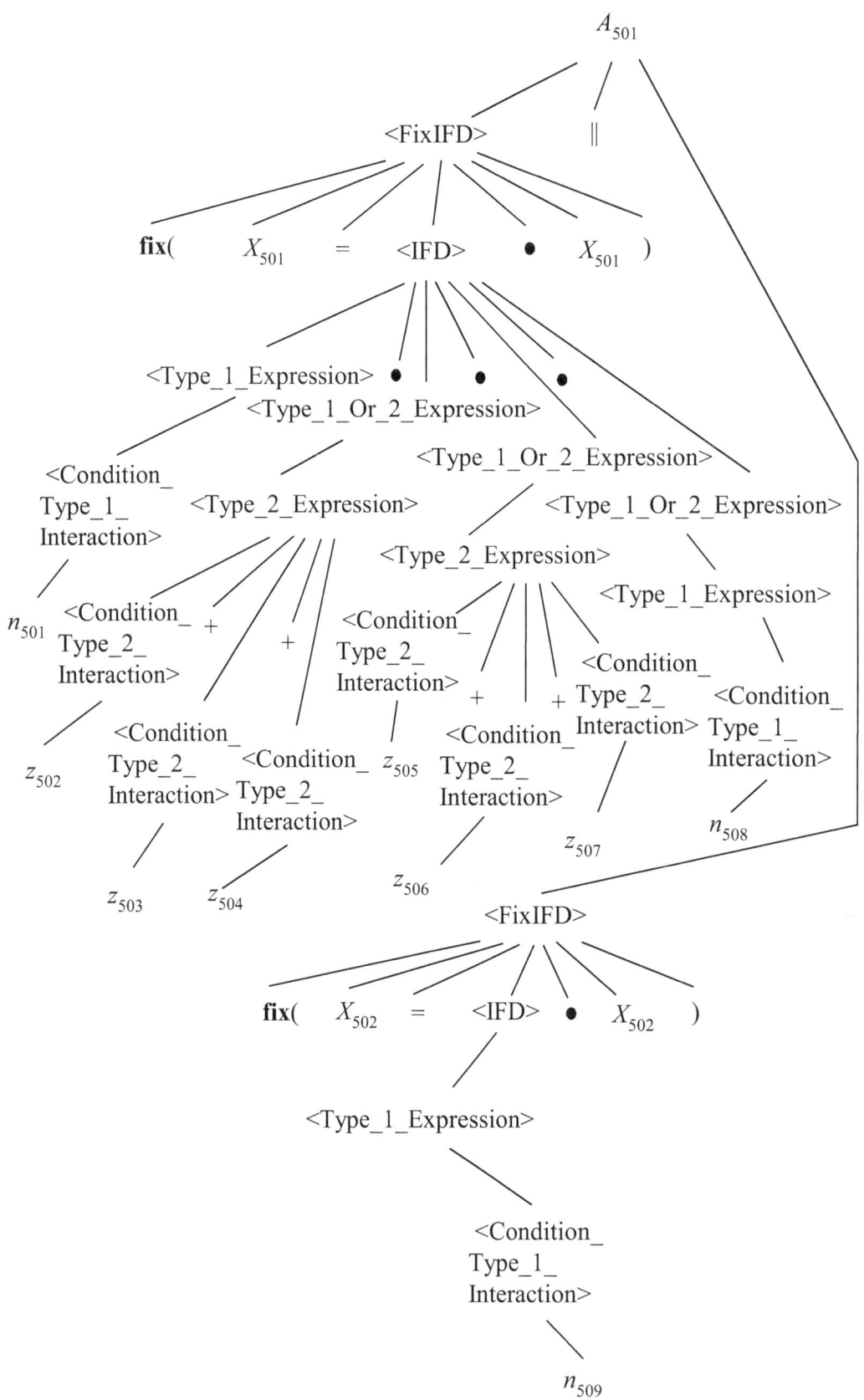

Figure 8-2 Backus-Naur Form Tree of the *Restaurant*'s Operation-Based Multi-Queue SBC Process

There are two IFDs in the operation-based multi-queue SBC process of the *Restaurant*. The first IFD is shown in Figure 8-3.

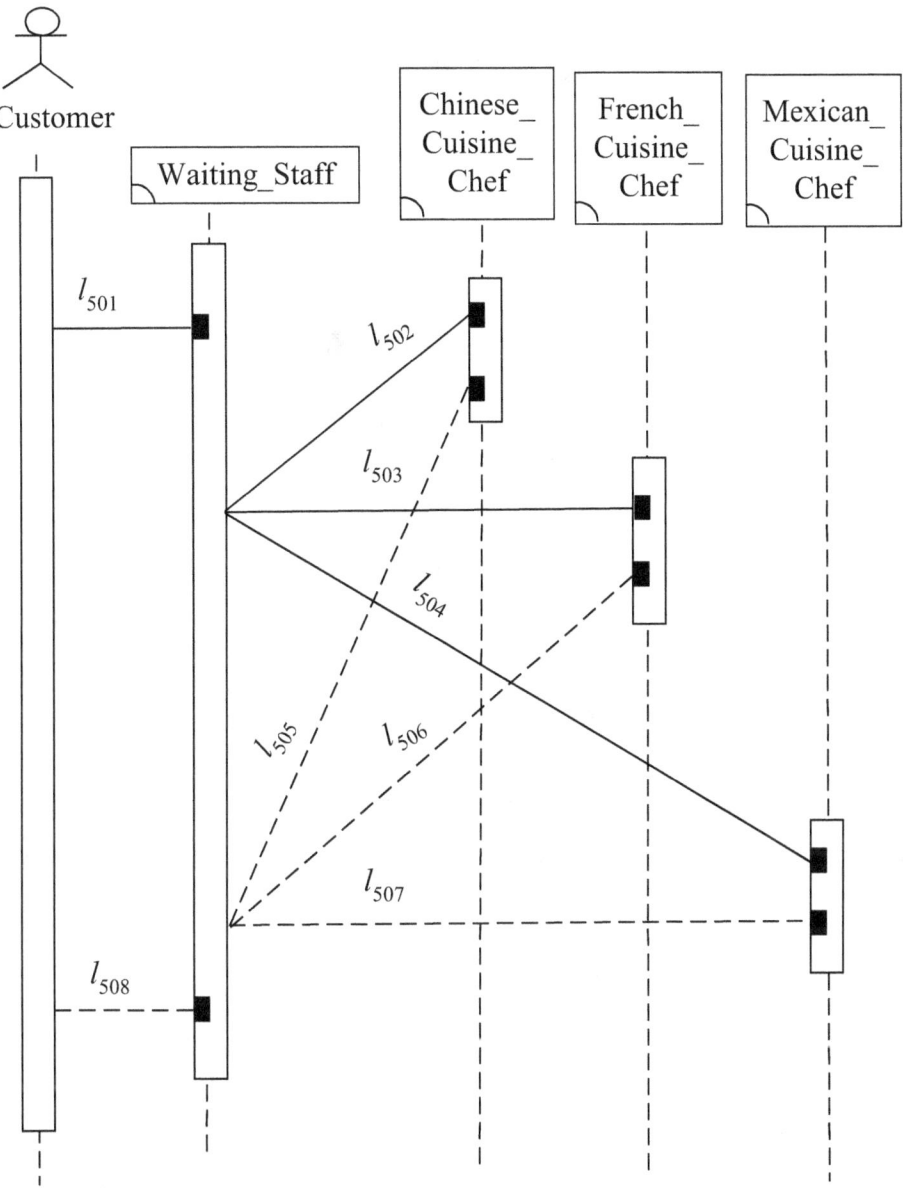

Figure 8-3　First IFD of the *Restaurant*

The second IFD of the operation-based multi-queue SBC process of the *Restaurant* is shown in Figure 8-4.

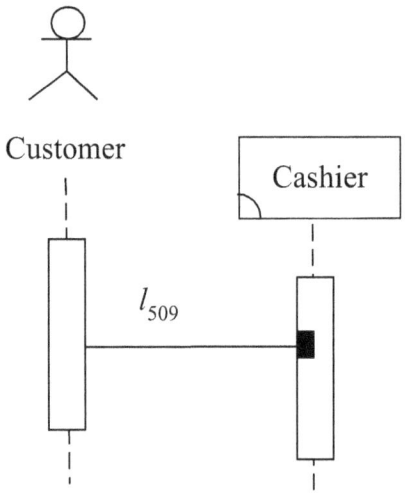

Figure 8-4   Second IFD of the *Restaurant*

8-2 Prefixes of the Restaurant

The n_{501} (operation-based condition type_1 interaction) prefix is defined as "<OPERATION_CALL, *Customer*, l_{501}, *Waiting_Staff*>", as shown in Figure 8-5.

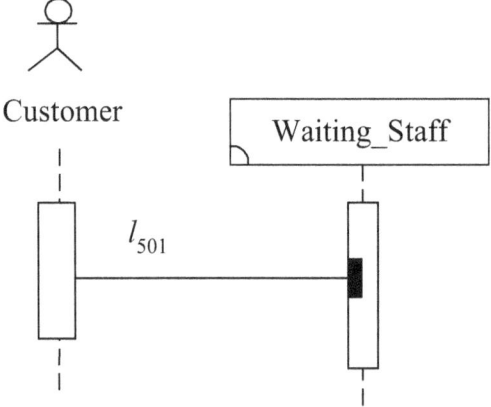

Figure 8-5   Prefix of n_{501}

The z_{502} (operation-based condition type_2 interaction) prefix is defined as "**if** Order = "Chinese" **then** <OPERATION_CALL, *Waiting_Staff*, l_{502}, *Chinese_Cuisine_Chef*>", as shown in Figure 8-6.

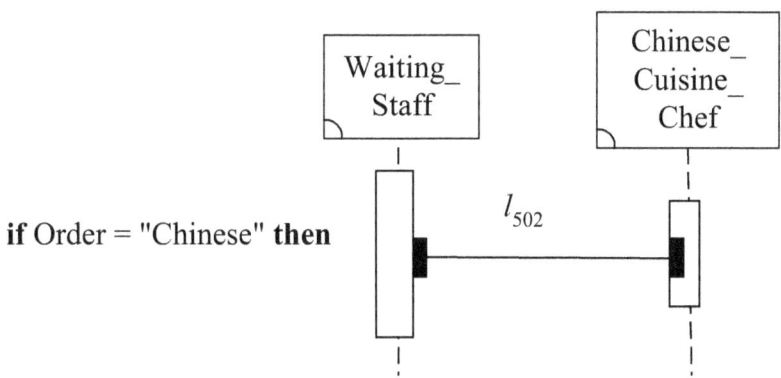

Figure 8-6   Prefix of z_{502}

The z_{503} (operation-based condition type_2 interaction) prefix is defined as "**if** Order = "French" **then** <OPERATION_CALL, *Waiting_Staff*, l_{503}, *French_Cuisine_Chef*>" as shown in Figure 8-7.

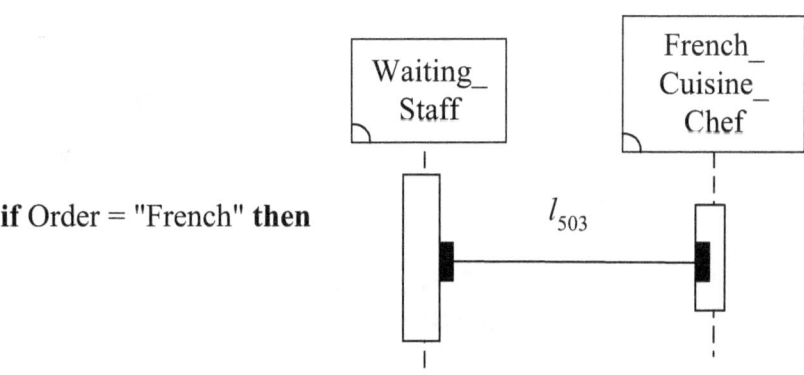

Figure 8-7   Prefix of z_{503}

The z_{504} (operation-based condition type_2 interaction) prefix is defined as "**if** Order = "Mexican" **then** <OPERATION_CALL, *Waiting_Staff*, l_{504}, *Mexican_Cuisine_Chef*>" as shown in Figure 8-8.

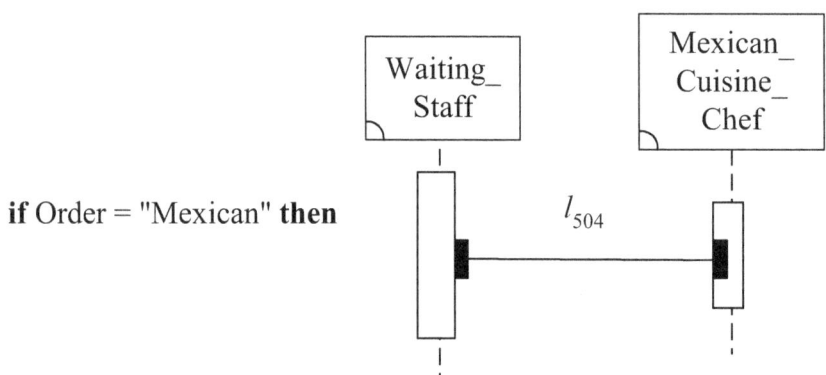

Figure 8-8   Prefix of z_{504}

The z_{505} (operation-based condition type_2 interaction) prefix is defined as "**if** Order = "Chinese" **then** <OPERATION_RETURN, *Waiting_Staff*, l_{505}, *Chinese_Cuisine_Chef*>" as shown in Figure 8-9.

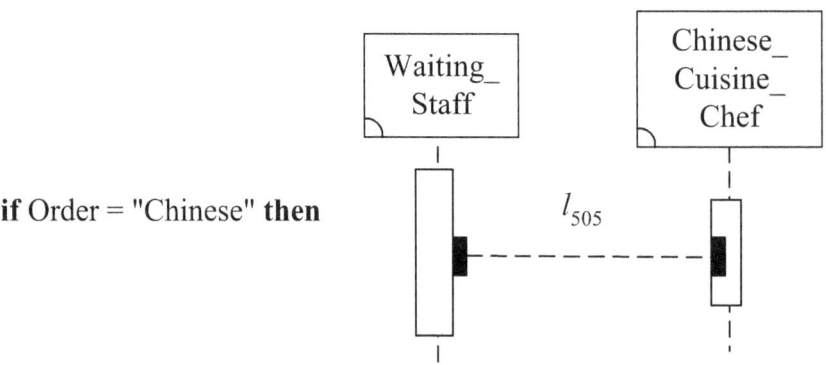

Figure 8-9   Prefix of z_{505}

The z_{506} (operation-based condition type_2 interaction) prefix is defined as "**if** Order = "French" **then** <OPERATION_RETURN, *Waiting_Staff*, l_{506}, *French_Cuisine_Chef*>" as shown in Figure 8-10.

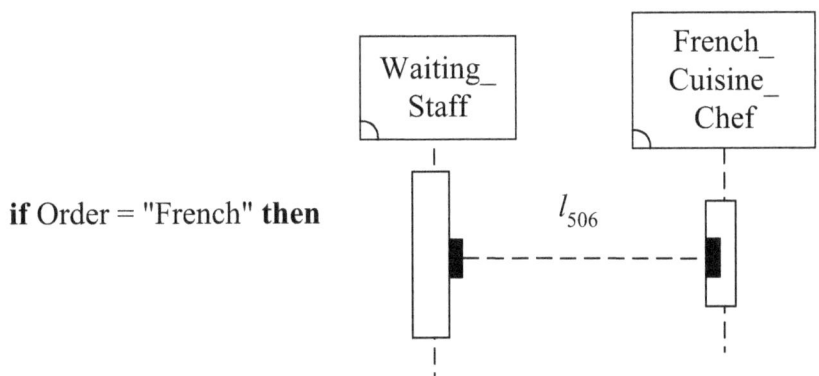

Figure 8-10 Prefix of z_{506}

The z_{507} (operation-based condition type_2 interaction) prefix is defined as "**if** Order = "Mexican" **then** <OPERATION_RETURN, *Waiting_Staff*, l_{507}, *Mexican_Cuisine_Chef*>" as shown in Figure 8-11.

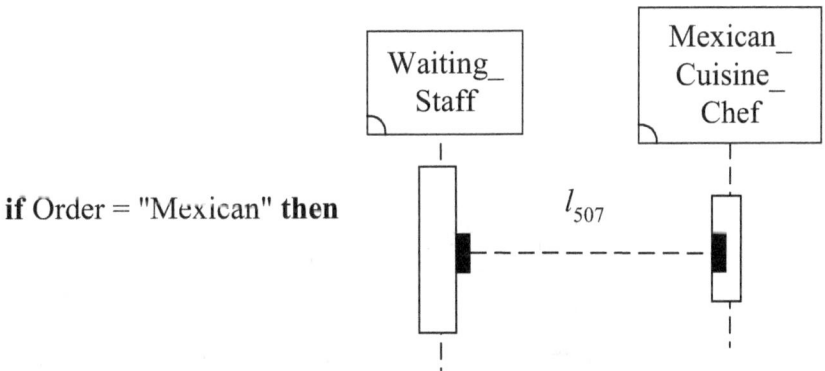

Figure 8-11 Prefix of z_{507}

The n_{508} (operation-based condition type_1 interaction) prefix is defined as "<OPERATION_RETURN, *Customer*, l_{508}, *Waiting_Staff*>", as shown in Figure 8-12.

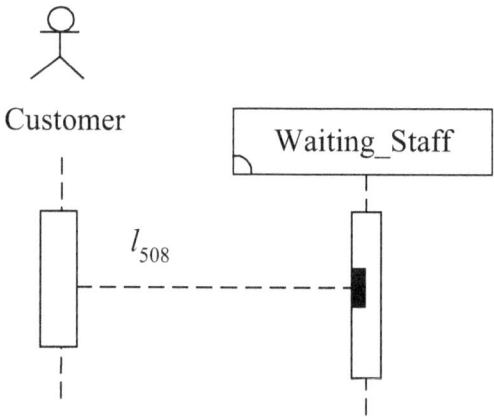

Figure 8-12   Prefix of n_{508}

The n_{509} (operation-based condition type_1 interaction) prefix is defined as "<OPERATION_CALL, *Customer*, l_{509}, *Cashier*>", as shown in Figure 8-13.

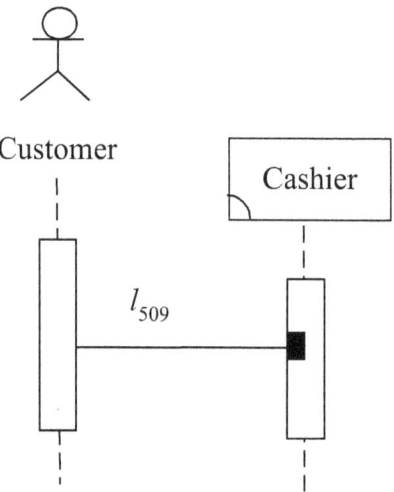

Figure 8-13   Prefix of n_{509}

Figure 8-14 shows all operation call or operation return formulas of the operation-based multi-queue SBC process of the *Restaurant*.

Entityname	Operation Call Formula	Operation Return Formula
l_{501}	Take_Order(In Customer_Request)	
l_{502}	Cook_Chinese_Food	
l_{503}	Cook_French_Food	
l_{504}	Cook_Mexican_Food	
l_{505}		Cook_Chinese_Food(Out Meal_1)
l_{506}		Cook_French_Food(Out Meal_2)
l_{507}		Cook_Mexican_Food(Out Meal_3)
l_{508}		Take_Order(Out Meal)
l_{509}	Pay_Bills	

Figure 8-14 Operation Call or Operation Return Formulas of the *Restaurant*'s Process

Figure 8-15 shows the primitive data type specification of the *Order* variable, the *Customer_Request* input parameter, and the *Meal, Meal_1, Meal_2, Meal_3* output parameters.

Parameter	Data Type	Instances
Order	Enumerated	Chinese, French, Mexican
Customer_Request	Enumerated	Chinese, French, Mexican
Meal	Enumerated food	Chinese food, French food, Mexican food
Meal_1	Food	Chinese food
Meal_2	Food	French food
Meal_3	Food	Mexican food

Figure 8-15   Primitive Data Type Specification

8-3 Process of the Restaurant

The following transition graph shows, in Figure 8-16, the semantics of P_{501}'s operation-based multi-queue SBC process.

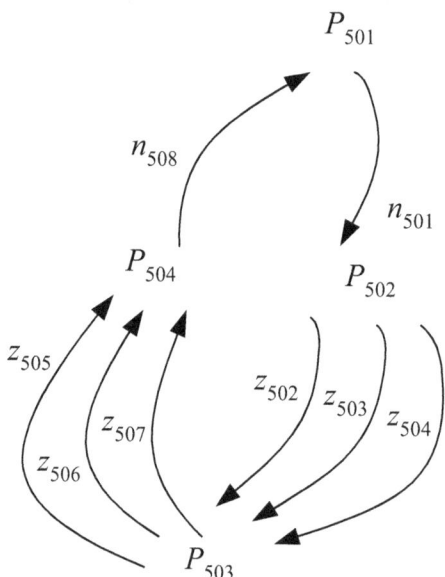

Figure 8-16   Transition graph of the P_{501} Process

In the transition graph of the P_{501}'s operation-based multi-queue SBC process, processes P_{501}, P_{502}, P_{503} and P_{504} are defined as in Figure 8-17.

$$P_{501} \stackrel{def}{=} n_{501} \bullet P_{502}$$
$$P_{502} \stackrel{def}{=} z_{502} \bullet P_{503} + z_{503} \bullet P_{503} + z_{504} \bullet P_{503}$$
$$P_{503} \stackrel{def}{=} z_{505} \bullet P_{504} + z_{506} \bullet P_{504} + z_{507} \bullet P_{504}$$
$$P_{504} \stackrel{def}{=} n_{508} \bullet P_{501}$$

Figure 8-17   Definition of Processes P_{501}, P_{502}, P_{503}, and P_{504}

The following transition graph shows, in Figure 8-18, the semantics of P_{505}'s operation-based multi-queue SBC process.

Figure 8-18   Transition graph of the P_{505} Process

In the transition graph of the P_{505}'s operation-based multi-queue SBC process, process P_{505} is defined as in Figure 8-19.

$$P_{505} \stackrel{def}{=} n_{509} \bullet P_{505}$$

Figure 8-19   Definition of Process P_{505}

The operation-based multi-queue SBC process of the *Restaurant*, A_{501}, is defined as "$P_{501} \| P_{505}$".

Chapter 9: Operation-Based Multi-Queue SBC Process of the Car

In this chapter, we use the operation-based multi-queue SBC process algebra to define the *Car* as shown in Figure 9-1.

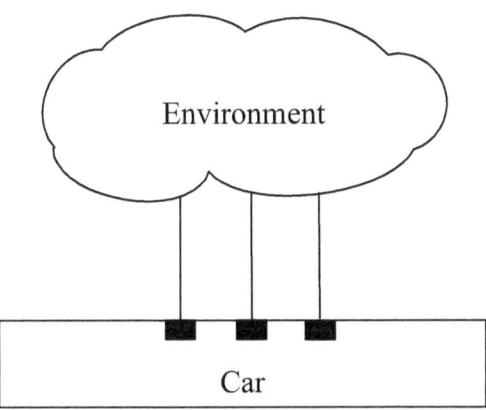

Figure 9-1 Systems Definition of the *Car*

9-1 BNF Tree of the Car

The operation-based multi-queue SBC process of the *Car*, A_{601}, is defined as "$\mathbf{fix}(X_{601}=n_{601}\bullet X_{601})\|\mathbf{fix}(X_{602}=n_{602}\bullet X_{602})\|\mathbf{fix}(X_{603}=n_{603}\bullet z_{604}\bullet(z_{605}+z_{606})\bullet X_{603})$".

We draw the operation-based multi-queue SBC process algebra Backus-Naur Form tree of A_{601} as shown in Figure 9-2.

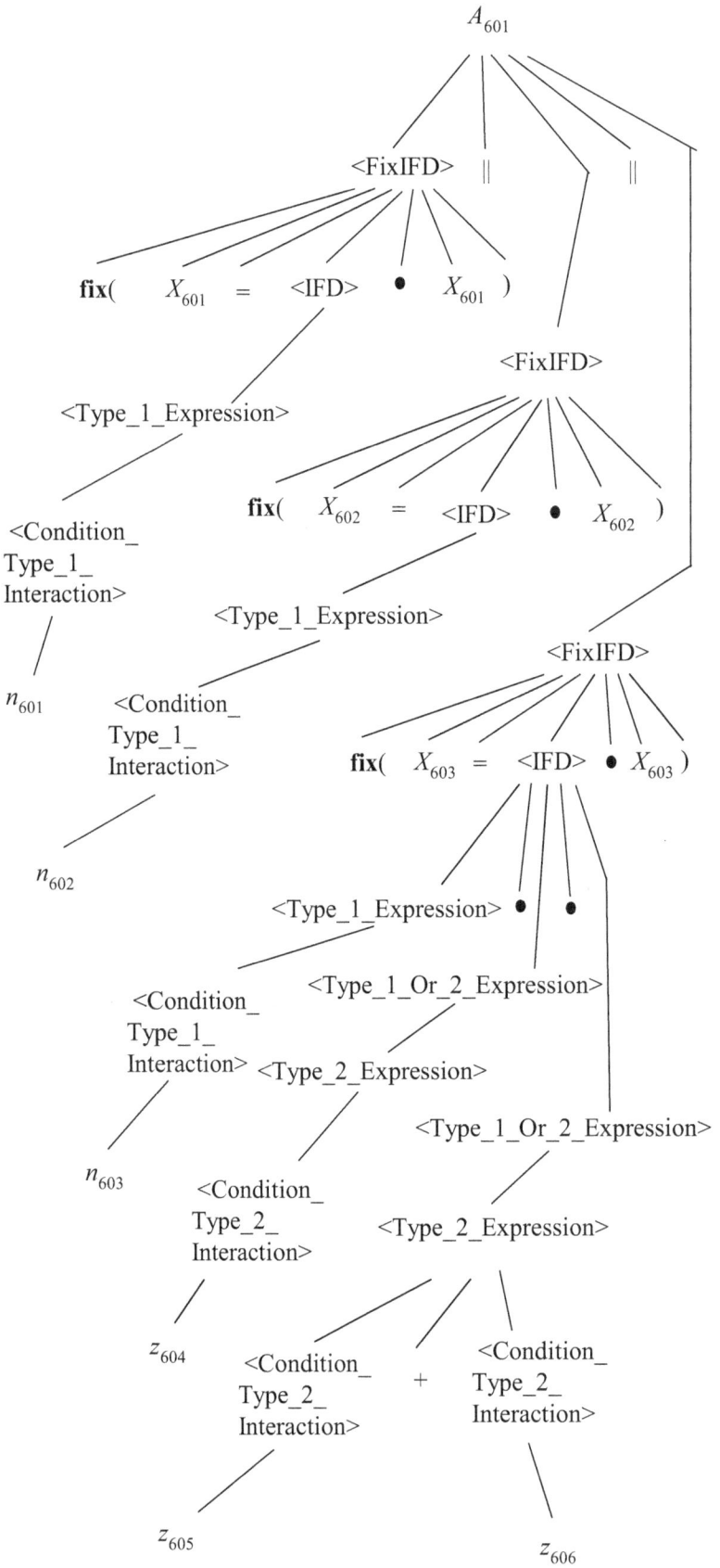

Figure 9-2 Backus-Naur Form Tree of the *Car*'s Operation-Based Multi-Queue SBC Process

There are three IFDs in the operation-based multi-queue SBC process of the *Car*. The first IFD is shown in Figure 9-3.

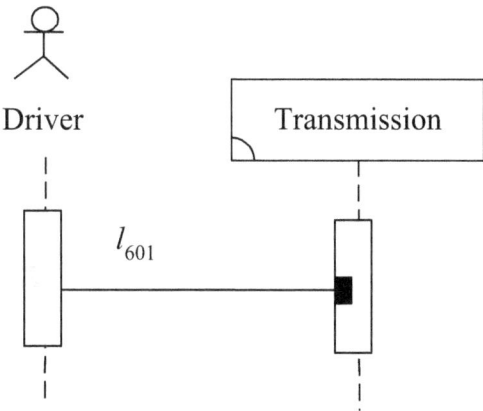

Figure 9-3   First IFD of the *Car*

The second IFD of the operation-based multi-queue SBC process of the *Car* is shown in Figure 9-4.

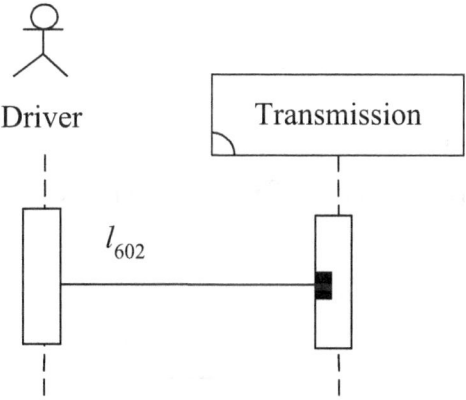

Figure 9-4   Second IFD of the *Car*

The third IFD of the operation-based multi-queue SBC process of the *Car* is shown in Figure 9-5.

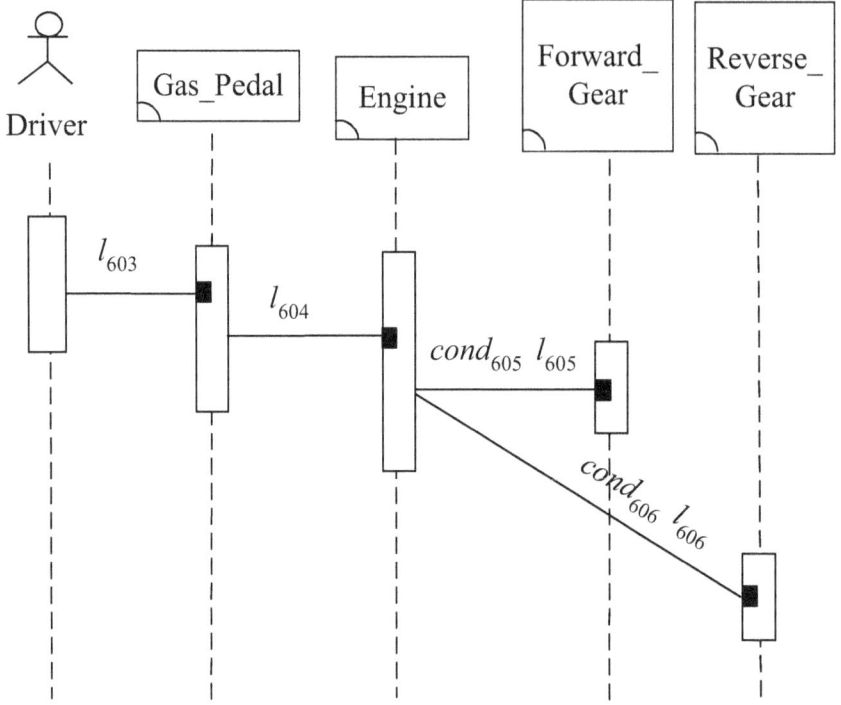

Figure 9-5　Third IFD of the *Car*

9-2 Prefixes of the Car

The n_{601} (operation-based condition type_1 interaction) prefix is defined as "<OPERATION_CALL, *Driver*, l_{601}, *Transmission*>", as shown in Figure 9-6.

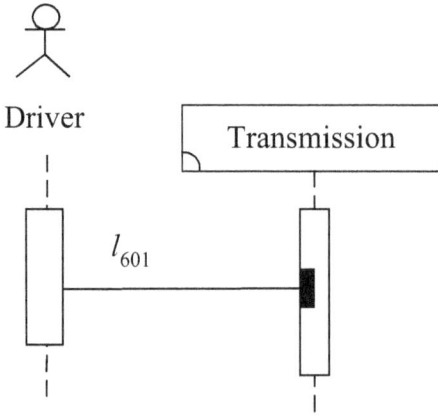

Figure 9-6　Prefix of n_{601}

The n_{602} (operation-based condition type_1 interaction) prefix is defined as "<OPERATION_CALL, *Driver*, l_{602}, *Transmission*>", as shown in Figure 9-7.

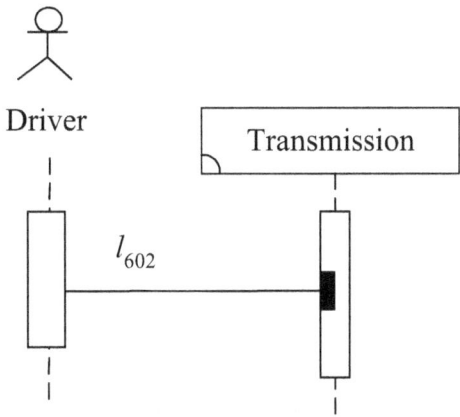

Figure 9-7 Prefix of n_{602}

The n_{603} (operation-based condition type_1 interaction) prefix is defined as "<OPERATION_CALL, *Driver*, l_{603}, *Gas_Pedal*>", as shown in Figure 9-8.

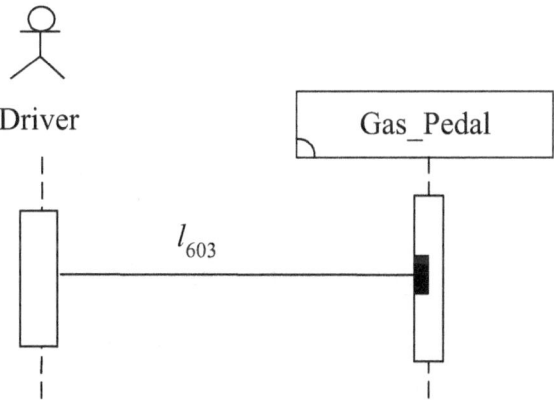

Figure 9-8 Prefix of n_{603}

The z_{604} (operation-based condition type_2 interaction) prefix is defined as "<OPERATION_CALL, *Gas_Pedal*, l_{604}, *Engine*>", as shown in Figure 9-9.

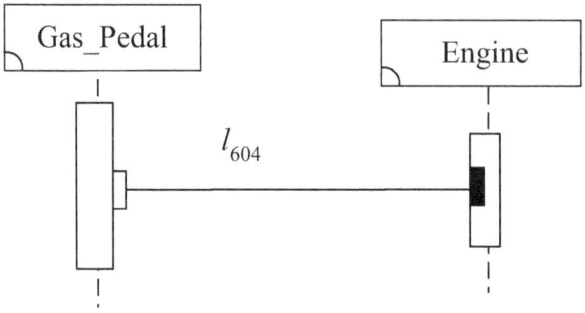

Figure 9-9   Prefix of z_{604}

The z_{605} (operation-based condition type_2 interaction) prefix is defined as "**if** Gear = "Forward" **then** <OPERATION_CALL, *Engine*, l_{605}, *Forward_Gear*>" as shown in Figure 9-10.

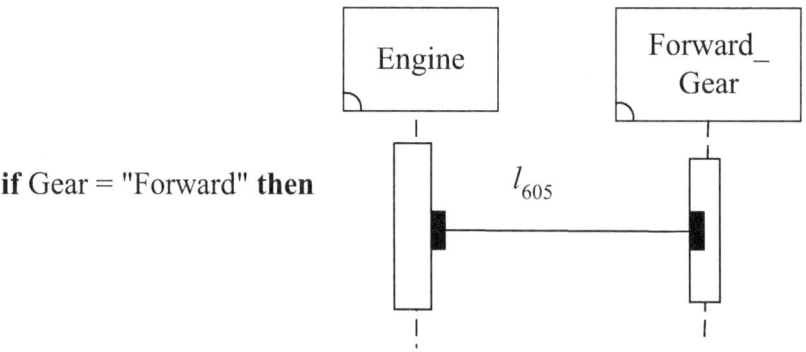

Figure 9-10   Prefix of z_{605}

The z_{606} (operation-based condition type_2 interaction) prefix is defined as "**if** Gear = "Reverse" **then** <OPERATION_CALL, *Engine*, l_{606}, *Reverse_Gear*>" as shown in Figure 9-11.

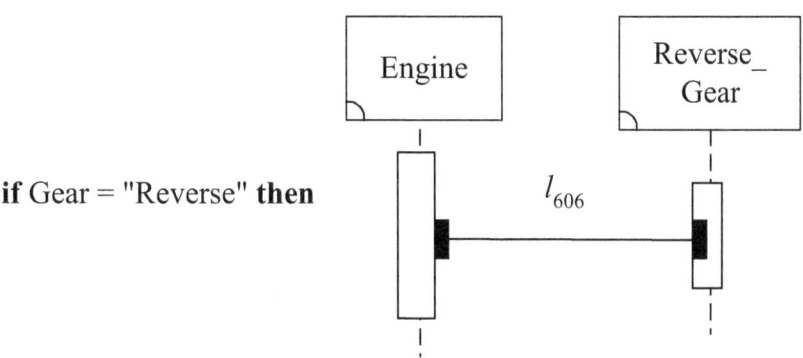

Figure 9-11   Prefix of z_{606}

Figure 9-12 shows all operation call or operation return formulas of the operation-based multi-queue SBC process of the *Car*.

Entity name	Operation Call Formula	Operation Return Formula
l_{601}	Push_Gear_Forward	
l_{602}	Push_Gear_Reverse	
l_{603}	Depress_Gas_Pedal	
l_{604}	Fuel_Supply	
l_{605}	Forward_Gear_Rotate	
l_{606}	Reverse_Gear_Rotate	

Figure 9-12   Operation Call or Operation Return Formulas of the *Car*'s Process

Figure 9-13 shows us the primitive data type specification of the *Gear* variable.

Parameter	Data Type	Instances
Gear	Enumerated	Forward, Reverse

Figure 9-13   Primitive Data Type Specification

9-3 Process of the Car

The following transition graph shows, in Figure 9-14, the semantics of P_{601}'s operation-based multi-queue SBC process.

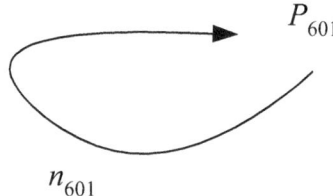

Figure 9-14   Transition graph of the P_{601} Process

In the transition graph of the P_{601}'s operation-based multi-queue SBC process, process P_{601} is defined as in Figure 9-15.

$$P_{601} \stackrel{\text{def}}{=} n_{601} \bullet P_{601}$$

Figure 9-15   Definition of Process P_{601}

The following transition graph shows, in Figure 9-16, the semantics of P_{602}'s operation-based multi-queue SBC process.

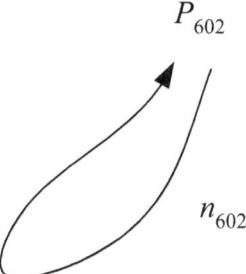

Figure 9-16   Transition graph of the P_{602} Process

In the transition graph of the P_{602}'s operation-based multi-queue SBC process, process P_{602} is defined as in Figure 9-17.

$$P_{602} \stackrel{\text{def}}{=\joinrel=} n_{602} \bullet P_{602}$$

Figure 9-17   Definition of Process P_{602}

The following transition graph shows, in Figure 9-18, the semantics of P_{603}'s operation-based multi-queue SBC process.

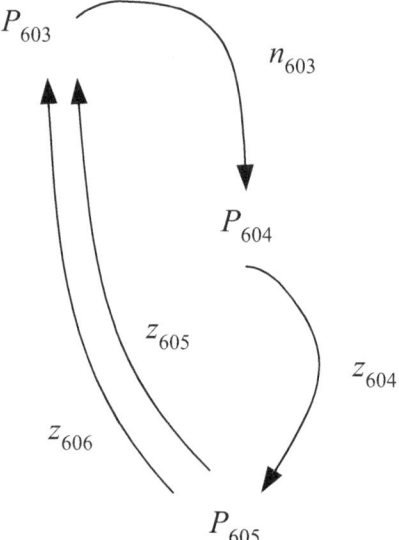

Figure 9-18 Transition graph of the P_{603} Process

In the transition graph of the P_{603}'s operation-based multi-queue SBC process, processes P_{603}, P_{604} and P_{605} are defined as in Figure 9-19.

$$P_{603} \stackrel{\text{def}}{=\!=} n_{603} \bullet P_{604}$$

$$P_{604} \stackrel{\text{def}}{=\!=} z_{604} \bullet P_{605}$$

$$P_{605} \stackrel{\text{def}}{=\!=} z_{605} \bullet P_{603} + z_{606} \bullet P_{603}$$

Figure 9-19 Definition of Processes P_{603}, P_{604}, and P_{605}

The operation-based multi-queue SBC process of the *Car*, A_{601}, is defined as "$P_{601} \| P_{602} \| P_{603}$".

APPENDIX A: Language Constructs of Operation-Based Multi-Queue SBC Process Algebra

(1) <System> ::= <FixIFD> {"||" <FixIFD>}

(2) <FixIFD> ::= "fix(" <Process_Variable>"="<IFD>
         "●" <Process_Variable> ")"

(3) <IFD> ::= <Type_1_Expression> {"●" <Type_1_Or_2_Expression>}

(4) <Type_1_Or_2_Expression> ::= <Type_1_Expression>
         | <Type_2_Expression>

(5) <Type_1_Expression> ::= <Condition_Type_1_Interaction>
         {"+" <Condition_Type_1_Interaction>}

(6) <Type_2 Expression> ::= <Condition_Type_2_Interaction>
         {"+" <Condition_Type_2_Interaction>}

APPENDIX B: Transitional Semantics of Operation-Based Multi-Queue SBC Process Algebra

Prefix	$$t \bullet E \xrightarrow{t} E$$
Sum_j	$$\dfrac{E_j \xrightarrow{t} E'_j}{\sum_{i \in I} E_i \xrightarrow{t} E'_j} (j \in I)$$
Parallel_1	$$\dfrac{E \xrightarrow{t} E'}{E \parallel F \xrightarrow{t} E' \parallel F}$$
Parallel_2	$$\dfrac{F \xrightarrow{t} F'}{E \parallel F \xrightarrow{t} E \parallel F'}$$
Recursion	$$\dfrac{E\{\mathbf{fix}(X=E)/X\} \xrightarrow{t} E'}{\mathbf{fix}(X=E) \xrightarrow{t} E'}$$
Constant	$$\dfrac{P \xrightarrow{t} P'}{A \xrightarrow{t} P'} \quad (A \stackrel{\text{def}}{=} P)$$

BIBLIOGRAPHY

[Acko68] Ackoff, R., "Toward a System of Systems Concepts," *Modern Systems Research for the Behavioral Scientist: A Sourcebook*, Aldine Publishing Company, 1968.

[Berg87] Bergstra, J. A. et al., "ACPτ: A Universal Axiom System for Process Specification," *CWI Quarterly* 15, 1987, pp. 3-23.

[Burd10] Burd, S. D., *Systems Architecture*, 6th Edition, Cengage Learning, 2010.

[Chao14a] Chao, W. S., *Systems Thingking 2.0: Architectural Thinking Using the SBC Architecture Description Language*, CreateSpace Independent Publishing Platform, 2014.

[Chao14b] Chao, W. S., *General Systems Theory 2.0: General Architectural Theory Using the SBC Architecture*, CreateSpace Independent Publishing Platform, 2014.

[Chao14c] Chao, W. S., *Systems Modeling and Architecting: Structure-Behavior Coalescence for Systems Architecture*, CreateSpace Independent Publishing Platform, 2014.

[Chao15a] Chao, W. S., *Variants of Interaction Flow Diagrams: The Structure-Behavior Coalescence Approach*, CreateSpace Independent Publishing Platform, 2015.

[Chao15b] Chao, W. S., *A Process Algebra For Systems Architecture: The Structure-Behavior Coalescence Approach*, CreateSpace Independent Publishing Platform, 2015.

[Chao15c] Chao, W. S., *An Observation Congruence Model For Systems Architecture: The Structure-Behavior Coalescence Approach*, CreateSpace Independent

Publishing Platform, 2015.

[Chao15d] Chao, W. S., *Variants of SBC Process Algebra: The Structure-Behavior Coalescence Approach*, CreateSpace Independent Publishing Platform, 2015.

[Chao15e] Chao, W. S., *Single-Queue SBC Process Algebra For Systems Architecture: The Structure-Behavior Coalescence Approach*, CreateSpace Independent Publishing Platform, 2015.

[Chao15f] Chao, W. S., *Multi-Queue SBC Process Algebra For Systems Architecture: The Structure-Behavior Coalescence Approach*, CreateSpace Independent Publishing Platform, 2015.

[Chao15g] Chao, W. S., *Single-Queue SBC Observation Congruence Model For Systems Architecture: The Structure-Behavior Coalescence Approach*, CreateSpace Independent Publishing Platform, 2015.

[Chao15h] Chao, W. S., *Multi-Queue SBC Observation Congruence Model For Systems Architecture: The Structure-Behavior Coalescence Approach*, CreateSpace Independent Publishing Platform, 2015.

[Chao16a] Chao, W. S., *System: Contemporary Concept, Definition, and Language*, CreateSpace Independent Publishing Platform, 2016.

[Chao16b] Chao, W. S., *Systems Architecture of Electronic Toll Collection Cloud Applications and Services IoT System*, CreateSpace Independent Publishing Platform, 2016.

[Chao16c] Chao, W. S., *Systems Architecture of Smart Healthcare Cloud Applications and Services IoT System*, CreateSpace Independent Publishing Platform, 2016.

[Chao16d] Chao, W. S., *Systems Architecture of Smart Healthcare Cloud Applications and Services IoT System*, CreateSpace Independent Publishing

Platform, 2016.

[Chao16e] Chao, W. S., *Systems Architecture of Ridesharing Sharing Economy Cloud Applications and Services IoT System*, CreateSpace Independent Publishing Platform, 2016.

[Chao16f] Chao, W. S., *Systems Architecture of Handy Helper Sharing Economy Cloud Applications and Services IoT System*, CreateSpace Independent Publishing Platform, 2016.

[Chao16g] Chao, W. S., *Systems Architecture of Vacation Rental Cleaning Sharing Economy Cloud Applications and Services IoT System*, CreateSpace Independent Publishing Platform, 2016.

[Chec99] Checkland, P., *Systems Thinking, Systems Practice: Includes a 30-Year Retrospective*, 1st Edition, Wiley, 1999.

[Craw15] Crawley, P. et al., *System Architecture: Strategy and Product Development for Complex Systems*, Prentice Hall, 2015.

[Dam06] Dam, S., *DoD Architecture Framework: A Guide to Applying System Engineering to Develop Integrated Executable Architectures*, BookSurge Publishing, 2006.

[Date03] Date, C. J., *An Introduction to Database Systems*, 8th Edition, Addison Wesley, 2003.

[Elma10] Elmasri, R., *Fundamentals of Database Systems*, 6th Edition, Addison Wesley, 2010.

[Frie11] Friedenthal, S., et al., *A Practical Guide to SysML: The Systems Modeling Language*, Morgan Kaufmann, 2nd Edition, 2011.

[Ghar11] Gharajedaghi, J., *Systems Thinking: Managing Chaos and Complexity: A Platform for Designing Business Architecture*, Morgan Kaufmann, 2011.

[Hoar85] Hoare, C. A. R., *Communicating Sequential Processes*, Prentice-Hall, 1985.

[Maie09] Maier, M. W., *The Art of Systems Architecting*, 3rd Edition, CRC Press, 2009.

[Mead08] Meadows, D. H., *Thinking in Systems: A Primer*, Chelsea Green Publishing, 2008.

[Miln89] Milner, R., *Communication and Concurrency*, Prentice-Hall, 1989.

[Miln99] Milner, R., *Communicating and Mobile Systems: the π-Calculus*, 1st Edition, Cambridge University Press, 1999.

[O'Rou03] O'Rourke, C. et al, *Enterprise Architecture Using the Zachman Framework*, 1st Edition, Course Technology, 2003.

[Putm00] Putman, J. R. et al., *Architecting with RM-ODP*, Prentice-Hall, 2000.

[Rayn09] Raynard, B., *TOGAF The Open Group Architecture Framework 100 Success Secrets*, Emereo Pty Ltd, 2009.

[Roza11] Rozanski, N. et al., *Software Systems Architecture: Working With Stakeholders Using Viewpoints and Perspectives*, 2nd Edition, Addison-Wesley Professional, 2011.

[Scho10] Scholl, C., *Functional Decomposition with Applications to FPGA Synthesis*, Springer, 2010.

[Toga08] The Open Group, *TOGAF Version 9 - A Manual (TOGAF Series)*, 9th Edition, Van Haren Publishing, 2008.

INDEX

A

ACP. *See* algebra of communicating processes
action, 35, 36, 37
    called, 36, 37
    called action, 36
    calling, 35, 36
    calling action, 35, 36
agent
    callee, 34
    caller, 34
algebra of communicating processes, 25

B

building block. *See* component

C

calculus of communicating systems, 25
callee, 34
caller, 34
CCS. *See* calculus of communicating systems
channel-based infinite-queue SBC process algebra, 26
channel-based multi-queue SBC process algebra, 26
channel-based single-queue SBC process algebra, 26
C-I-SBC-PA. *See* channel-based infinite-queue SBC process algebra
C-M-SBC-PA. *See* channel-based multi-queue SBC process algebra
communicating sequential processes, 25
communication. *See* interaction
component, 44
    callee, 36
    caller, 34
CSP. *See* communicating sequential processes
C-S-SBC-PA. *See* channel-based single-queue SBC process algebra

E

entity. *See* component
external environment, 44

F

functional decomposition, 22, 24

G

generalized SBC process algebra, 25
G-SBC-PA. *See* generalized SBC process algebra

H

handshake. *See* interaction

I

IFD. *See* interaction flow diagram, *See* interaction flow diagram, *See* interaction flow diagram
interaction, 24
    operation call interaction, 34
    operation return interaction, 34
interaction flow diagram, 54, 55, 85, 86, 97

internal interaction, 38
    operation-based, 38

N

non-aggregated system. *See* component

O

object. *See* component

O-I-SBC-PA. *See* operation-based infinite-queue SBC process algebra

O-M-SBC-PA. *See* operation-based multi-queue SBC process algebra

operation, 33

operation call formula, 39, 40, 41, 45, 46, 49

operation call interaction, 34

operation formula, 34

operation return formula, 39, 40, 41, 45, 46, 49

operation return interaction, 34

operation-based action, 40

operation-based infinite-queue SBC process algebra, 26

operation-based interaction
    operation call, 34
    operation return, 35

operation-based multi-queue SBC process algebra, 26

operation-based port, 39

operation-based single-queue SBC process algebra, 26

O-S-SBC-PA. *See* operation-based single-queue SBC process algebra

P

parallel composition, 27

part. *See* component

port, 34, 35, 36, 37
    called, 36, 37
    calling, 34, 35

prefix, 50, 51, 52, 53, 57, 58, 59, 60, 61, 62, 63, 64, 65, 66, 67, 68, 69, 70, 71, 72, 73, 76

process algebra, 11, 25
    algebra of communicating processes, 25
    calculus of communicating systems, 25
    channel-based infinite-queue SBC process algebra, 26
    channel-based multi-queue SBC process algebra, 26
    channel-based single-queue SBC process algebra, 26
    communicating sequential processes, 25
    generalized SBC process algebra, 25
    operation-based infinite-queue SBC process algebra, 26
    operation-based multi-queue SBC process algebra, 26
    operation-based single-queue SBC process algebra, 26

R

recursion, 28

replication, 28

S

SBC. *See* structure-behavior coalescence

sequentialization, 27

structural decomposition, 22, 24

structure-behavior coalescence, 43, 44

summation, 27

system, 17, 18
    abstract system, 18
    boundary, 19

concrete system, 17
environment, 19
isolated system, 21
notional system, 18
open system, 20
physical system, 17
real system, 17
virtual system, 18
systems, 43, 44

systems behavior, 43, 44
systems definition, 11, 17, 26
    systems definition 1.0, 21
    systems definition 2.0, 23
systems structure, 43, 44

T

transition graph, 92, 93, 102, 103

www.ingramcontent.com/pod-product-compliance
Lightning Source LLC
Chambersburg PA
CBHW082340220526
45470CB00008B/2578